Lecture Notes in Computer Science   14671

Founding Editors

Gerhard Goos
Juris Hartmanis

The series Lecture Notes in Computer Science (LNCS), including its subseries Lecture Notes in Artificial Intelligence (LNAI) and Lecture Notes in Bioinformatics (LNBI), has established itself as a medium for the publication of new developments in computer science and information technology research, teaching, and education.

LNCS enjoys close cooperation with the computer science R & D community, the series counts many renowned academics among its volume editors and paper authors, and collaborates with prestigious societies. Its mission is to serve this international community by providing an invaluable service, mainly focused on the publication of conference and workshop proceedings and postproceedings. LNCS commenced publication in 1973.

Megan Dewar · Bogumił Kamiński ·
Daniel Kaszyński · Łukasz Kraiński ·
Paweł Prałat · François Théberge ·
Małgorzata Wrzosek
Editors

Modelling and Mining Networks

19th International Workshop, WAW 2024
Warsaw, Poland, June 3–6, 2024
Proceedings

 Springer

Editors
Megan Dewar
Tutte Institute for Mathematics
and Computing
Ottawa, ON, Canada

Daniel Kaszyński
SGH Warsaw School of Economics
Warsaw, Poland

Paweł Prałat
Toronto Metropolitan University
Toronto, ON, Canada

Małgorzata Wrzosek
SGH Warsaw School of Economics
Warsaw, Poland

Bogumił Kamiński
SGH Warsaw School of Economics
Warsaw, Poland

Łukasz Kraiński
SGH Warsaw School of Economics
Warsaw, Poland

François Théberge
Tutte Institute for Mathematics
and Computing
Ottawa, ON, Canada

ISSN 0302-9743   ISSN 1611-3349 (electronic)
Lecture Notes in Computer Science
ISBN 978-3-031-59204-1   ISBN 978-3-031-59205-8 (eBook)
https://doi.org/10.1007/978-3-031-59205-8

This Springer imprint is published by the registered company Springer Nature Switzerland AG
The registered company address is: Gewerbestrasse 11, 6330 Cham, Switzerland

Paper in this product is recyclable.

Preface

The *19th Workshop on Modelling and Mining Networks (WAW 2024)* was held at the SGH Warsaw School of Economics, Warsaw, Poland (June 3–6, 2024). This is an annual meeting, providing opportunities for researchers in network science to interact and to exchange research ideas. We do hope that the event was an effective venue for the dissemination of new results and for fostering research collaboration.

Virtually every human-technology interaction, or sensor network, generates observations that are in some relation with each other. As a result, many data science problems can be viewed as a study of some properties of complex networks in which nodes represent the entities that are being studied and edges represent relations between these entities. Such networks are often large-scale, decentralized, and evolve dynamically over time. Modelling and mining complex networks in order to understand the principles governing the organization and the behaviour of such networks is crucial for a broad range of fields of study, including information and social sciences, economics, biology, and neuroscience.

The aim of the *19th Workshop on Modelling and Mining Networks (WAW 2024)* was to further the understanding of networks that arise in theoretical as well as applied domains. The goal was also to stimulate the development of high-performance and scalable algorithms that exploit these networks. The workshop welcomed researchers who are working on graph-theoretic and algorithmic aspects of networks represented as graphs or hypergraphs and other higher–order structures.

This volume contains the papers accepted and presented during the workshop. Each submission was carefully reviewed by the members of the Programme Committee. Papers were submitted and reviewed using the EasyChair online system. The committee members decided to accept 12 papers.

June 2024

Megan Dewar
Bogumił Kamiński
Daniel Kaszyński
Łukasz Kraiński
Paweł Prałat
François Théberge
Małgorzata Wrzosek

Organization

General Chairs

Andrei Z. Broder   Google Research, USA
Fan Chung Graham   University of California San Diego, USA

Organizing Committee

Megan Dewar   Tutte Institute for Mathematics and Computing,
     Canada
Bogumił Kamiński   SGH Warsaw School of Economics, Poland
Daniel Kaszyński   SGH Warsaw School of Economics, Poland
Lukasz Kraiński   SGH Warsaw School of Economics, Poland
Paweł Prałat   Toronto Metropolitan University, Canada
François Théberge   Tutte Institute for Mathematics and Computing,
     Canada
Małgorzata Wrzosek   SGH Warsaw School of Economics, Poland

Sponsoring Institutions

NAWA – The Polish National Agency for Academic Exchange
SGH Warsaw School of Economics
Toronto Metropolitan University
Tutte Institute for Mathematics and Computing

Program Committee

Konstantin Avratchenkov   Inria, France
Leman Akoglu   Carnegie Mellon University, USA
Mindaugas Bloznelis   Vilnius University, Lithuania
Paolo Boldi   University of Milan, Italy
Anthony Bonato   Toronto Metropolitan University, Canada
Ulrik Brandes   ETH Zürich, Switzerland
Fan Chung Graham   UC San Diego, USA
Collin Cooper   King's College London, UK

Contents

Subgraph Counts in Random Clustering Graphs 1
  Fan Chung and Nicholas Sieger

Self-similarity of Communities of the ABCD Model 17
  Jordan Barrett, Bogumił Kamiński, Paweł Prałat, and François Théberge

A Simple Model of Influence: Details and Variants of Dynamics 32
  Colin Cooper, Nan Kang, Tomasz Radzik, and Ngoc Vu

Impact of Market Design and Trading Network Structure on Market
Efficiency 47
  Nick Arnosti, Bogumił Kamiński, Paweł Prałat, and Mateusz Zawisza

Network Embedding Exploration Tool (NEExT) 65
  Ashkan Dehghan, Paweł Prałat, and François Théberge

Efficient Computation of K-Edge Connected Components: An Empirical
Analysis 80
  Hanieh Sadri, Venkatesh Srinivasan, and Alex Thomo

The Directed Age-Dependent Random Connection Model with Arc
Reciprocity 97
  Lukas Lüchtrath and Christian Mönch

How to Cool a Graph 115
  Anthony Bonato, Holden Milne, Trent G. Marbach, and Teddy Mishura

Distributed Averaging for Accuracy Prediction in Networked Systems 130
  Christel Sirocchi and Alessandro Bogliolo

Towards Graph Clustering for Distributed Computing Environments 146
  Przemysław Szufel

HypergraphRepository: A Community-Driven and Interactive
Hypernetwork Data Collection 159
  Alessia Antelmi, Daniele De Vinco, and Carmine Spagnuolo

x   Contents

Clique Counts for Network Similarity 174
    Anthony Bonato and Zhiyuan Zhang

Author Index 185

Subgraph Counts in Random Clustering Graphs

Fan Chung and Nicholas Sieger[(✉)]

University of California, San Diego, La Jolla, CA 92037, USA
fan@ucsd.edu, nsieger@ucsd.edu

Abstract. We analyze subgraph counts in random clustering graphs for general degree distributions. Building on the prior work, we weaken the assumptions required to derive our previous results and exactly determine the asymptotics of subgraph counts in a random clustering graphs under mild conditions. As an application, we analyze the clustering coefficient and cycle counts in random clustering graphs.

Keywords: Random graphs · clustering coefficient · scale-free networks · Chung-Lu model

1 Introduction

As can be seen in examples as diverse as collaboration graphs of mathematicians, clause-variable incidence graphs of SAT instances, and protein interaction networks, many real-world networks exhibit a *clustering effect* where two nodes are more likely to be adjacent if they share a common neighbor. These same examples also show that real-world networks have heterogeneous degree sequences, where there are many low degree vertices and a few high degree vertices. There has much interest in producing random graph models which reproduce clustering behavior and heterogeneous degree sequences, with models built on geometric assumptions [1,6,7,10], planted communities [4,5,9], and random dot products [11] amongst others.

In previous work [8], we introduced a model which captures both clustering and heterogeneous degree sequences without additional geometric assumptions. The model builds on the Chung-Lu model introduced in [2] by adding additional edges in local neighborhoods according to a parameter γ associated with the clustering effect.

In the classical Erdős-Rényi model $G(n,p)$, subgraph counts depend only on a single parameter p, whereas in random clustering graphs with expected degrees defined by a vector $\mathbf{d}$, subgraph counts depend on higher moments of $\mathbf{d}$. In the study of random clustering graphs in [8], we determined subgraph counts by controlling all of the higher moments of $\mathbf{d}$. Here we show that control of just two moments suffices to determine the clustering effect in a random clustering graph. Furthermore, we precisely compute the subgraph count of an arbitrary graph in a random clustering graph.

M. Dewar et al. (Eds.): WAW 2024, LNCS 14671, pp. 1–16, 2024.
https://doi.org/10.1007/978-3-031-59205-8_1

Nowadays, we face a myriad of intricate questions in an evolving network as we strive to predict its behavior. Often, we wish to predict global behavior of the network by examining local structures. At the heart of any local structure lies the question of determining subgraph counts. In this work we focus on one local invariant, the clustering coefficient, and determine its value for a random clustering graph. Our result goes further, however, and determines general subgraph counts in networks which can be modeled by random clustering graphs. It is our goal that this model and the methods provided here will serve as a useful tool for many other challenging problems.

The paper is organized as follows. We state our notation and give the definitions of the random graph models needed in Sect. 2. In section Sect. 3 we prove our main result Theorem 1. We showed in [8] that the subgraph count of H in a random clustering graph G is determined by the subgraph counts of the *extension configurations* of H in the Chung-Lu model, but left open the question of which specific extension configurations control subgraph counts of H. In Theorem 1, we show that one extension configuration dominates the subgraph counts of H for a wide range of vectors $\mathbf{d}$ which greatly simplifies the probabilistic and combinatorial analysis of random clustering graphs. In Sect. 4, we use Theorem 1 to determine the expected subgraph counts of arbitrary H in Theorem 2 and then show that subgraph counts are concentrated around their means in Theorem 3. Finally, we apply our results to specific subgraphs in Sect. 5, in particular cycle counts and the determination of the clustering coefficient.

2 Preliminaries

All graphs will be simple, undirected, and labeled unless explicitly noted otherwise. We write $E(G)$ and $V(G)$ for the edge and vertex sets of a graph G and write $e(G)$ and $v(G)$ for the number of edges and number of vertices respectively.

We consider labeled subgraph counts defined as follows. A *graph homomorphism* of a graph H to a graph G is a map $\phi : V(H) \to V(G)$ such that $uv \in E(H) \implies \phi(u)\phi(v) \in E(G)$. The set of graph homomorphisms from H to G will be denoted by $\mathrm{Hom}(H,G)$, and the number of homomorphisms will be denoted by $\hom(H,G)$. We also write $H \curvearrowright_\phi G$ to denote the event that the map $\phi : V(H) \to V(G)$ is a graph homomorphism of H into G.

We will also consider graph homomorphisms which are restricted in the following sense. For a graph H, a subset $S \subseteq V(H)$, and an map $\psi : S \to V(G)$, let $\mathrm{Hom}_\psi(H,G)$ be the set of graph homomorphisms $\phi : V(H) \to V(G)$ such that $\psi(v) = \phi(v)$ for every $v \in S$. Similarly, let $\hom_\psi(H,G) = \left|\mathrm{Hom}_\psi(H,G)\right|$.

If μ is a probability distribution on a set Ω, and $P(x)$ is a proposition on the variable $x \in \Omega$, then $\mathbb{P}_{x \leftarrow \mu}\left[P(x)\right]$ will denote the probability that $P(x)$ holds when x is drawn from the distribution μ. Similarly, for a random variable $R(x)$ on Ω, we write $\mathbb{E}_{x \leftarrow \mu}\left[R(x)\right]$ to denote the expected value of R when x is distributed according to μ. Whenever we write the expectation or probability over a set, such as $\mathbb{E}_{x \in \Omega}$, the expectation or probability is taken with respect to the uniform distribution on Ω.

2.1   Volumes

The *kth-order volume* of a graph G is

$$\text{vol}_k(G) = \text{vol}(G, k) = \sum_{v \in V(G)} \deg_G(v)^k$$

and for a vector $\mathbf{d} \in \mathbb{R}^n$ we define

$$\text{Vol}_k(\mathbf{d}) = \text{Vol}(\mathbf{d}, k) = \sum_{i \in [n]} \mathbf{d}_i^k.$$

We write $\text{Vol}(\mathbf{d}) = \text{Vol}(\mathbf{d}, 1)$. Throughout the paper we will assume that $\mathbf{d}_i \geq 1$ for every $i \in [n]$. With this assumption, we have that $\text{Vol}(\mathbf{d}, 0) = n$. The *k-order average degree* of a graph G, denoted $\delta_k(G)$, is defined by

$$\delta_k(G) = \frac{\text{vol}_k(G)}{\text{vol}_{k-1}(G)}$$

and likewise the *kth-order average degree* of a vector $\mathbf{d}$, denoted $\delta_k(\mathbf{d})$ is

$$\delta_k(\mathbf{d}) = \frac{\text{Vol}_k(\mathbf{d})}{\text{Vol}_{k-1}(\mathbf{d})}.$$

Note that $\delta_1(\mathbf{d})$ is simply the average of the entries of $\mathbf{d}$.

Remark 1. For a fixed vector $\mathbf{d}$ with maximum entry $\mathbf{d}_{\max}$, the following hold:

(i) $1 \leq \delta_i(\mathbf{d}) \leq \mathbf{d}_{\max}$ for every $i \geq 1$.
(ii) For $k \geq l$, $\delta_k(\mathbf{d}) \geq \delta_l(\mathbf{d})$ by Holder's inequality.
(iii) $\text{Vol}_k(\mathbf{d}) = n \prod_{i=1}^{k} \delta_k(\mathbf{d})$ and in particular $\text{Vol}_1(\mathbf{d}) = n\delta_1(\mathbf{d})$.

We will write δ_i for $\delta_i(\mathbf{d})$ when the vector $\mathbf{d}$ is clear from context.

2.2   The Chung-Lu Model

For completeness, we state the definition of the Chung-lu model:

Definition 1. *Fix a nonnegative vector $\mathbf{d}$ such that $\mathbf{d}_{\max}^2 \leq \text{Vol}(\mathbf{d})$. The Chung-Lu model, denoted $\mathcal{G}(\mathbf{d})$, is the distribution on graphs constructed as follows. For each pair of distinct vertices u, v the edge uv appears in $\mathcal{G}(\mathbf{d})$ independently of all other edges with probability $\frac{\mathbf{d}_u \mathbf{d}_v}{\text{Vol}(\mathbf{d})}$.*

We will need the following lemma regarding subgraph counts in the Chung-Lu model.

Lemma 1. *[8] Fix a graph H. If $\mathbf{d} \in \mathbb{R}^n$ is a nonnegative vector, then the subgraph count of H in the Chung-Lu model $\mathcal{G}(\mathbf{d})$ is determined by the degree sequence of H and the kth-order volumes of $\mathbf{d}$, i.e.,*

$$\mathbb{E}_{G \hookleftarrow \mathcal{G}(\mathbf{d})} \left[\text{hom}(H, G) \right] = \frac{\prod_{v \in H} \text{Vol}(\mathbf{d}, \deg_H(v))}{\text{Vol}(\mathbf{d})^{e(H)}} \tag{1}$$

Proof. By definition,

$$
\mathbb{E}_{G \hookleftarrow \mathcal{G}(\mathbf{d})} \left[\hom(H, G) \right] = \sum_{\psi : H \to [n]} \mathbb{P}_{G \hookleftarrow \mathcal{G}(\mathbf{d})} \left[\forall uv \in E(H) : \psi(u)\psi(v) \in E(G) \right]
$$

$$
= \sum_{\psi : H \to [n]} \prod_{uv \in E(H)} \mathbb{P}_{G \hookleftarrow \mathcal{G}(\mathbf{d})} \left[\psi(u)\psi(v) \in E(G) \right]
$$

$$
= \sum_{\psi : H \to [n]} \prod_{uv \in E(H)} \frac{\mathbf{d}_{\phi(u)} \mathbf{d}_{\phi(v)}}{\mathrm{Vol}(\mathbf{d})}
$$

$$
= \frac{\prod_{v \in H} \mathrm{Vol}(\mathbf{d}, \deg_H(v))}{\mathrm{Vol}(\mathbf{d})^{e(H)}}
$$

□

2.3   Random Clustering Graphs

Our work focuses on the following random graph model introduced in [8].

Definition 2. *Fix a nonnegative vector* $\mathbf{d}$ *and* $\gamma \in [0, 1]$. *The random clustering graph, denoted* $\mathcal{C}(\gamma, \mathbf{d})$, *is the distribution on graphs constructed as follows:*

1. *Generate a graph* G_0 *drawn from* $\mathcal{G}(\mathbf{d})$.
2. *For each path with edges* uv, vw *in* G_0, *add the edge* uw *independently with probability* γ.

3   Extension Configurations

Definition 3. *An extension configuration for a graph* H *is a graph* X *on the vertex set* $V(H) \sqcup \{v_1, \ldots, v_l\}$ *such that*

- *For every* $uv \in H \setminus X$, *there is a vertex* $x \in X$ *such that the edges* ux, vx *are in* X.
- X *is edge-minimal with respect to the first condition.*
- H *has no isolated vertices.*

The *external vertices* of an extension configuration X of H, denoted $\mathrm{Ex}(X)$, are the vertices $V(X) \setminus V(H)$. Let $\mathrm{ex}(X)$ denote $|\mathrm{Ex}(X)|$. Let $\mathrm{Ext}(H)$ denote the set of extension configurations of a graph H. Note that H may be disconnected, which will be used in Theorem 3. See Fig. 1 for the example of the extension configurations of K_3.

There are several facts regarding extension configurations which we will use in Theorem 1.

Lemma 2. *Fix a graph* H. *If* X *is an extension configuration of* H, *then*

1. *If* H *is connected, so is* X.
2. *Every external vertex in* $x \in \mathrm{Ex}(X)$ *has degree at least 2.*

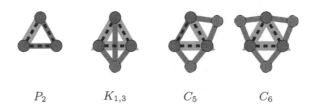

$$P_2 \qquad K_{1,3} \qquad C_5 \qquad C_6$$

Fig. 1. The four extension configurations (in thick solid red) for K_3 with the edges of K_3 in dashed black. (Color figure online)

3. *There are no edges between external vertices in X.*
4. *Every $v \in V(H)$ has at most $\deg_H(v)$ external neighbors in X.*

One extension configuration will be of particular importance. The *star extension* on H, denoted $K_{1,v(H)}$, is the extension configuration of H with one external vertex adjacent to all vertices in H.

In [8], we showed that the subgraph count of H in a random clustering graph is given by a linear combination of subgraph counts of $X \in \mathrm{Ext}(H)$ as subgraphs in the Chung-Lu model. In the remainder of this section, we prove that the star extension dominates the sum in H under certain assumptions on $\mathbf{d}$. First we relate the subgraph count of a graph to the subgraph count of a spanning tree.

Lemma 3. *Fix a connected graph H. If T is a spanning tree of H, then*

$$\frac{\mathbb{E}_{G \leftarrow \mathcal{G}(\mathbf{d})} \big[\mathrm{hom}(H, G)\big]}{\mathbb{E}_{G \leftarrow \mathcal{G}(\mathbf{d})} \big[\mathrm{hom}(T, G)\big]} \leq \left(\frac{\mathbf{d}_{\max}^2}{\mathrm{Vol}(\mathbf{d})} \right)^{e(H) - v(H) + 1}. \tag{2}$$

Proof. It H is already a tree, we finished. Otherwise, let uv be an edge not in T, and let $H' = H - uv$. By Lemma 1 and the fact that $e(H') = e(H) - 1$,

$$\frac{\mathbb{E}_{G \leftarrow \mathcal{G}(\mathbf{d})} \big[\mathrm{hom}(H, G)\big]}{\mathbb{E}_{G \leftarrow \mathcal{G}(\mathbf{d})} \big[\mathrm{hom}(H', G)\big]} = \frac{1}{\mathrm{Vol}(\mathbf{d})} \frac{\prod_{v \in H} \mathrm{Vol}(\mathbf{d}, \deg_H(v))}{\prod_{v \in H'} \mathrm{Vol}(\mathbf{d}, \deg_{H'}(v))}. \tag{3}$$

As the only vertices whose degrees change are u and v, $\deg_{H'}(u) = \deg_H(u) - 1$, and $\deg_{H'}(v) = \deg_H(v) - 1$,

$$\frac{\mathbb{E}_{G \leftarrow \mathcal{G}(\mathbf{d})} \big[\mathrm{hom}(H, G)\big]}{\mathbb{E}_{G \leftarrow \mathcal{G}(\mathbf{d})} \big[\mathrm{hom}(H', G)\big]} = \left(\frac{1}{\mathrm{Vol}(\mathbf{d})} \right) \left(\frac{\mathrm{Vol}(\mathbf{d}, \deg_H(u)) \, \mathrm{Vol}(\mathbf{d}, \deg_H(v))}{\mathrm{Vol}(\mathbf{d}, \deg_H(u) - 1) \, \mathrm{Vol}(\mathbf{d}, \deg_H(v) - 1)} \right)$$

$$\leq \frac{\delta_{\deg_H(u)} \delta_{\deg_H(v)}}{\mathrm{Vol}(\mathbf{d})} \leq \frac{\mathbf{d}_{\max}^2}{\mathrm{Vol}(\mathbf{d})}.$$

The statement follows by induction. $\square$

The following is a key lemma in the proof of Theorem 1.

Lemma 4. *Fix a graph H, a connected extension X of H, and a spanning tree T of X. If H has maximum degree Δ, then*

$$e(X) - v(X) + 1 \geq \frac{1}{2}\left(l - v(H) + \frac{\text{ex}(X)}{\Delta}\right) \tag{4}$$

where l denotes the number of leaves in T.

Proof. Let L be the set of leaves in T which are not leaves in X. By definition, each vertex in L is incident to an edge in X which is not an edge in T. As these edges may join two vertices in L, it follows that $e(X) - v(X) + 1 \geq \frac{1}{2}|L|$.

If L_0 is the set of leaves in X, we have $l = |L| + |L_0|$ as no vertex in L is a leaf in X. To conclude the lemma, we upper bound $|L_0|$. By Lemma 2, no external vertex in X a leaf, and thus $L_0 \subseteq V(H)$ is in L_0. Let $A = \{v \in V(H) : \deg_X(v) > 1\}$. Since X is connected and there is no edge between two external vertices, each external vertex is adjacent to at least one vertex in A. By Lemma 2 each vertex $v \in A$ has at most $\deg_H(v) \leq \Delta$ external neighbors. Thus $\Delta|A| \geq \text{ex}(X)$. It follows that

$$|L| = l - |L_0| = l - (v(H) - |A|) \geq l - \left(v(H) - \frac{\text{ex}(X)}{\Delta}\right) \tag{5}$$

and the lemma follows. $\square$

Next, we need a technical lemma. We define $\ell(H, i) = |\{v \in V(H) : \deg_H(v) \geq i\}|$.

Lemma 5. *Let T be a tree. If T has l leaves, then*

$$\sum_{i>2} \ell(T, i) = l - 2. \tag{6}$$

Proof. Let L denote the set of leaf vertices, and let $\overline{L}$ denote the non-leaf vertices. Observe that

$$\sum_{v \in \overline{L}} (\deg_T(v) - 2) = \sum_{v \in \overline{L}} (\deg_T(v) - 2) = \sum_{v \in \overline{L}} \sum_{i>2} [\deg_T(v) \geq i] = \sum_{i>2} \ell(T, i). \tag{7}$$

By the handshaking lemma,

$$\sum_{v \in \overline{L}} (\deg_T(v) - 2) = (2e(T) - l) - 2|\overline{L}| = (2(v(T) - 1) - l) - 2(v(T) - l) = l - 2 \tag{8}$$

and the lemma follows. $\square$

We can now prove Theorem 1

Theorem 1. *Fix a graph H and a connected extension configuration X of H. If X is not the star extension on H, then X is asymptotically dominated by the star extension provided that $\delta_2(\mathbf{d})^{1+\Delta(H)} = o\left(\delta_3(\mathbf{d})\right)$ and $\mathbf{d}_{\max} = o(n^{\frac{1}{4}})$.*

Proof. Let T be a spanning tree of X and let l be the number of leaves in T. Let X' denote the star extension on H. Throughout the proof, we write δ_i for $\delta_i(\mathbf{d})$. By Lemma 1,

$$\frac{\mathbb{E}_{G\leftarrow\mathcal{G}(\mathbf{d})}\left[\hom(T,G)\right]}{\mathbb{E}_{G\leftarrow\mathcal{G}(\mathbf{d})}\left[\hom(X',G)\right]} = \frac{1}{\text{Vol}(\mathbf{d})^{e(T)-e(X')}}\frac{\prod_{v\in T}\text{Vol}(\mathbf{d},\deg_T(v))}{\prod_{v\in X'}\text{Vol}(\mathbf{d},\deg_{X'}(v))}$$

$$= \frac{n^{v(T)}\prod_{v\in T}\left(\prod_{i=1}^{\deg_T(v)}\delta_i\right)}{\left(n^{e(T)-e(X')}\delta_1^{e(T)-e(X')}\right)\left(n^{v(H)+1}\delta_1^{v(H)}\left(\prod_{i=1}^{v(H)}\delta_i\right)\right)}.$$

Since $e(T) = v(H) + \text{ex}(X) - 1$, $v(T) = v(H) + \text{ex}(X)$, and $e(X') = v(H) - 1$,

$$\frac{\mathbb{E}_{G\leftarrow\mathcal{G}(\mathbf{d})}\left[\hom(T,G)\right]}{\mathbb{E}_{G\leftarrow\mathcal{G}(\mathbf{d})}\left[\hom(X',G)\right]} = \frac{\prod_{i=1}^{\Delta(T)}\delta_i^{\ell(T,i)}}{\delta_1^{v(T)}\prod_{i=2}^{v(H)}\delta_i}. \tag{9}$$

Since $\ell(T,1) = v(T)$,

$$\frac{\mathbb{E}_{G\leftarrow\mathcal{G}(\mathbf{d})}\left[\hom(T,G)\right]}{\mathbb{E}_{G\leftarrow\mathcal{G}(\mathbf{d})}\left[\hom(X',G)\right]} = \frac{\prod_{i=2}^{\Delta(T)}\delta_i^{\ell(T,i)}}{\prod_{i=2}^{v(H)}\delta_i}. \tag{10}$$

Observe that $\ell(T,1) - \ell(T,2) = l$. We have

$$\frac{\mathbb{E}_{G\leftarrow\mathcal{G}(\mathbf{d})}\left[\hom(T,G)\right]}{\mathbb{E}_{G\leftarrow\mathcal{G}(\mathbf{d})}\left[\hom(X',G)\right]} = \delta_2^{v(T)-l-1}\left(\frac{\prod_{i=3}^{\Delta(T)}\delta_i^{\ell(T,i)}}{\prod_{i=3}^{v(H)}\delta_i}\right). \tag{11}$$

By Lemma 5, $\sum_{i>2}\ell(T,i) = l - 2$ and we consider the sequence consisting of $\ell(T,i)$ copies of each δ_i for $3 \leq i \leq \Delta(T)$ in increasing order. If $l \leq v(H)$, this sequence is dominated by $\delta_3, \delta_4, \dots, \delta_{v(H)}$ since $\delta_i \geq \delta_j$ for $i \geq j$. Therefore,

$$\frac{\prod_{i=3}^{\Delta(T)}\delta_i^{\ell(T,i)}}{\prod_{i=3}^{v(H)}\delta_i} \leq \frac{1}{\prod_{i=l+1}^{v(H)}\delta_i} \leq \frac{1}{\delta_3^{v(H)-l}}. \tag{12}$$

If $l > v(H)$, we can remove the $l - v(H)$ additional terms as the end of the sequence and apply the same argument to show that

$$\frac{\prod_{i=3}^{\Delta(T)}\delta_i^{\ell(T,i)}}{\prod_{i=3}^{v(H)}\delta_i} \leq \prod_{i=v(H)+1}^{l}\delta_i \leq \text{d}_{\max}^{l-v(H)}. \tag{13}$$

Thus if $l \leq v(H)$,

$$\frac{\mathbb{E}_{G\leftarrow\mathcal{G}(\mathbf{d})}\left[\hom(T,G)\right]}{\mathbb{E}_{G\leftarrow\mathcal{G}(\mathbf{d})}\left[\hom(X',G)\right]} \leq \frac{\delta_2^{v(T)-l-1}}{\delta_3^{v(H)-l}}$$

$$= o\left(\delta_3^{\frac{1}{\Delta(H)+1}(v(T)-l-1)-(v(H)-l)}\right) \tag{14}$$

$$= o\left(\delta_3^{\frac{\text{ex}(X)-1}{\Delta(H)+1}-\left(1-\frac{1}{\Delta(H)+1}\right)(v(H)-l)}\right) \tag{15}$$

where we use our assumption on δ_2 in Eq. (14). If $l > v(H)$, we have

$$\frac{\mathbb{E}_{G \leftarrow \mathcal{G}(\mathbf{d})}\left[\hom(T,G)\right]}{\mathbb{E}_{G \leftarrow \mathcal{G}(\mathbf{d})}\left[\hom(X',G)\right]} \leq \delta_2^{v(T)-l-1}\mathbf{d}_{\max}^{l-v(H)}$$

$$\leq o\left(\delta_3^{\frac{v(T)-l-1}{\Delta(H)+1}}\mathbf{d}_{\max}^{l-v(H)}\right) \tag{16}$$

$$\leq o\left(\mathbf{d}_{\max}^{\frac{v(T)-l-1}{\Delta(H)+1}-(l-v(H))}\right) \tag{17}$$

$$= o\left(\mathbf{d}_{\max}^{\frac{\mathrm{ex}(X)-1}{\Delta(H)+1}-\left(1-\frac{1}{\Delta(H)+1}\right)(l-v(H))}\right) \tag{18}$$

where we use our assumption on δ_2 in Eq. (16) and the fact that T can have no more than $v(T) - 1$ leaves in Eq. (17).

As $\mathbf{d}_{\max} \leq n^{\frac{1}{4}}$ and $\mathrm{Vol}(\mathbf{d}) \geq n$, $\frac{\mathbf{d}_{\max}^2}{\mathrm{Vol}(\mathbf{d})} = o\left(n^{-\frac{1}{2}}\right)$. By Lemma 3

$$\frac{\mathbb{E}_{G \leftarrow \mathcal{G}(\mathbf{d})}\left[\hom(X,G)\right]}{\mathbb{E}_{G \leftarrow \mathcal{G}(\mathbf{d})}\left[\hom(T,G)\right]} \leq o\left(\left(n^{-\frac{1}{2}}\right)^{e(X)-v(X)+1}\right). \tag{19}$$

Since X is connected, $e(X) - v(X) + 1 \geq 0$. By Lemma 4, we have

$$e(X) - v(X) + 1 \geq \max\left\{0, \frac{1}{2}\left(l - v(H) + \frac{\mathrm{ex}(X)}{\Delta(H)}\right)\right\}. \tag{20}$$

With Eqs. (15), (18) and (20) in hand, we complete the proof by considering the following three cases based on the number of leaves in T.

Case (i) $l \leq v(H) - \frac{\mathrm{ex}(X)}{\Delta(H)}$. Equation (15), (19) and (20) imply that

$$\frac{\mathbb{E}_{G \leftarrow \mathcal{G}(\mathbf{d})}\left[\hom(X,G)\right]}{\mathbb{E}_{G \leftarrow \mathcal{G}(\mathbf{d})}\left[\hom(X',G)\right]} \leq \frac{\mathbb{E}_{G \leftarrow \mathcal{G}(\mathbf{d})}\left[\hom(T,G)\right]}{\mathbb{E}_{G \leftarrow \mathcal{G}(\mathbf{d})}\left[\hom(X',G)\right]}$$

$$= o\left(\delta_3^{\frac{\mathrm{ex}(X)-1}{\Delta(H)+1}-\left(1-\frac{1}{\Delta(H)+1}\right)(v(H)-l)}\right).$$

As $v(H) - l \geq \frac{\mathrm{ex}(X)}{\Delta(H)}$, we have

$$\frac{\mathbb{E}_{G \leftarrow \mathcal{G}(\mathbf{d})}\left[\hom(X,G)\right]}{\mathbb{E}_{G \leftarrow \mathcal{G}(\mathbf{d})}\left[\hom(X',G)\right]} \leq o\left(\delta_3^{\frac{\mathrm{ex}(X)-1}{\Delta(H)+1}-\left(1-\frac{1}{\Delta(H)+1}\right)\frac{\mathrm{ex}(X)}{\Delta(H)}}\right)$$

$$\leq o\left(\delta_3^{\frac{-1}{\Delta(H)+1}}\right)$$

$$= o(1).$$

Case (ii) $l > v(H)$. Equation (19) and (20) imply that

$$\frac{\mathbb{E}_{G \leftarrow \mathcal{G}(\mathbf{d})}\left[\hom(X, G)\right]}{\mathbb{E}_{G \leftarrow \mathcal{G}(\mathbf{d})}\left[\hom(T, G)\right]} \leq o\left(n^{\frac{-1}{4}\left(l - v(H) + \frac{\text{ex}(X)}{\Delta(H)}\right)}\right).$$

Combining Eq. (18) and $\mathbf{d}_{\max} = o\left(n^{\frac{1}{4}}\right)$, we have

$$\frac{\mathbb{E}_{G \leftarrow \mathcal{G}(\mathbf{d})}\left[\hom(T, G)\right]}{\mathbb{E}_{G \leftarrow \mathcal{G}(\mathbf{d})}\left[\hom(X', G)\right]} \leq o\left(\mathbf{d}_{\max}^{\frac{\text{ex}(X)-1}{\Delta(H)+1} - \left(1 - \frac{1}{\Delta(H)+1}\right)(l - v(H))}\right)$$

$$\leq o\left(n^{\frac{1}{4}\left(\frac{\text{ex}(X)-1}{\Delta(H)+1} - \left(1 - \frac{1}{\Delta(H)+1}\right)(l - v(H))\right)}\right).$$

Thus,

$$\frac{\mathbb{E}_{G \leftarrow \mathcal{G}(\mathbf{d})}\left[\hom(X, G)\right]}{\mathbb{E}_{G \leftarrow \mathcal{G}(\mathbf{d})}\left[\hom(X', G)\right]} \leq o\left(n^{\frac{1}{4}\left(\frac{\text{ex}(X)-1}{\Delta(H)+1} - \left(1 - \frac{1}{\Delta(H)+1}\right)(l - v(H))\right) - \frac{1}{4}\left(l - v(H) + \frac{\text{ex}(X)}{\Delta(H)}\right)}\right)$$

$$\leq o\left(n^{\frac{1}{4}\frac{\text{ex}(X)-1}{\Delta(H)+1} - \frac{1}{4}\frac{\text{ex}(X)}{\Delta(H)}}\right)$$

$$= o(1).$$

where we use $l > v(H)$ in the second line.

Case (iii) $v(H) - \text{ex}(X)/\Delta(H) < l \leq v(H)$. Consider $r = l - v(H) + \frac{\text{ex}(X)}{\Delta(H)}$, and note that $r > 0$. Equation (19) and (20) imply that

$$\frac{\mathbb{E}_{G \leftarrow \mathcal{G}(\mathbf{d})}\left[\hom(X, G)\right]}{\mathbb{E}_{G \leftarrow \mathcal{G}(\mathbf{d})}\left[\hom(T, G)\right]} \leq o\left(n^{\frac{1}{4}\left(v(H) - l - \frac{\text{ex}(X)}{\Delta(H)}\right)}\right) = o\left(n^{\frac{-r}{4}}\right). \quad (21)$$

By Eq. (15),

$$\frac{\mathbb{E}_{G \leftarrow \mathcal{G}(\mathbf{d})}\left[\hom(T, G)\right]}{\mathbb{E}_{G \leftarrow \mathcal{G}(\mathbf{d})}\left[\hom(X', G)\right]} \leq \delta_3^{\frac{\text{ex}(X)-1}{\Delta(H)+1} - \left(1 - \frac{1}{\Delta(H)+1}\right)(v(H) - l)}$$

$$\leq n^{\frac{1}{4}\left(\frac{\text{ex}(X)-1}{\Delta(H)+1} - \left(1 - \frac{1}{\Delta(H)+1}\right)(v(H) - l)\right)}.$$

where we use $\delta_3 \leq \mathbf{d}_{\max}$ and $\mathbf{d}_{\max} = o(n^{\frac{1}{4}})$ in the final line. As $v(H) - l = \frac{\text{ex}(X)}{\Delta(H)} - r$, we have

$$\frac{\mathbb{E}_{G \leftarrow \mathcal{G}(\mathbf{d})}\left[\hom(T, G)\right]}{\mathbb{E}_{G \leftarrow \mathcal{G}(\mathbf{d})}\left[\hom(X', G)\right]} \leq n^{\frac{1}{4}\left(\frac{\text{ex}(X)-1}{\Delta(H)+1} - \left(1 - \frac{1}{\Delta(H)+1}\right)\left(\frac{\text{ex}(X)}{\Delta(H)} - r\right)\right)}$$

$$\leq n^{\frac{1}{4}\left(1 - \frac{1}{\Delta(H)+1}\right) r}. \quad (22)$$

Combining Eqs. (21) and (22) we find that

$$\frac{\mathbb{E}_{G \leftarrow \mathcal{G}(\mathbf{d})}\left[\hom(X,G)\right]}{\mathbb{E}_{G \leftarrow \mathcal{G}(\mathbf{d})}\left[\hom(X',G)\right]} = o\left(n^{\frac{1}{4}\left(1-\frac{1}{\Delta(H)+1}\right)r-\frac{r}{4}}\right)$$

$$= o(1).$$

as $r > 0$.

$\square$

Remark 2. When $\mathbf{d}_{\max}$ exceeds $n^{\frac{1}{4}}$, Theorem 1 can still be used to predict behavior on a large subgraph, namely the subgraph of vertices of degree at most $n^{\frac{1}{4}}$.

Remark 3. We note that random clustering graphs only exhibit clustering behavior when $\delta_2 = o(\delta_3)$, and thus our assumption that $\delta_2^{1+\Delta(H)} = o(\delta_3)$ quantifies how far these two quantities must be. As our desired applications (Theorems 4 and 5) have $\Delta(H) = 2$, the requirement that $\delta_2^3 = o(\delta_3)$ uses small constants, and in particular, does not grow with the number of vertices in H.

We can extend Theorem 1 to disconnected graphs.

Corollary 1. *Let H be a graph and let X be an extension configuration of H. Fix $\gamma \in [0,1]$. If $\mathbf{d} \in \mathbb{R}^n$ is a vector such that $\delta_2(\mathbf{d})^{\Delta(H)+1} = o(\delta_3(\mathbf{d}))$ and $\mathbf{d}_{\max} = o(\min\{n^{\frac{1}{4}}, \mathrm{Vol}(\mathbf{d}, v(H))^{\frac{1}{v(H)}}\})$ and X which is not the star extension on each connected component of H, then*

$$\frac{\mathbb{E}_{G \leftarrow \mathcal{G}(\mathbf{d})}\left[\hom(X,G_0)\right]}{\mathbb{E}_{G \leftarrow \mathcal{G}(\mathbf{d})}\left[\hom(X',G_0)\right]} = o(1) \tag{23}$$

where X' is the star extension on each connected component of H.

Proof. We induct on the number of components. If H is connected, X is connected by Lemma 2 and Theorem 1 proves the base case. Otherwise, assume that H has at least two connected components. If X is connected and not the star extension, then Theorem 1 implies that $\frac{\mathbb{E}_{G \leftarrow \mathcal{G}(\mathbf{d})}\left[\hom(X,G_0)\right]}{\mathbb{E}_{G \leftarrow \mathcal{G}(\mathbf{d})}\left[\hom(K_{1,v(H)},G_0)\right]} = o(1)$. Let C be a connected component of H, and let X_C be the extension configuration of H with a star extension on C and a star extension on $V(H) - C$. We then have

$$\frac{\mathbb{E}_{G \leftarrow \mathcal{G}(\mathbf{d})}\left[\hom(K_{1,v(H)},G_0)\right]}{\mathbb{E}_{G \leftarrow \mathcal{G}(\mathbf{d})}\left[\hom(X_C,G_0)\right]} = \frac{\mathrm{Vol}(\mathbf{d}, v(H))}{\mathrm{Vol}(\mathbf{d}, |C|)\,\mathrm{Vol}(\mathbf{d}, v(H) - |C|)} \leq \frac{\mathbf{d}_{\max}^{|C|}}{\mathrm{Vol}(\mathbf{d}, |C|)} = o(1) \tag{24}$$

by Lemma 1 and the assumption that $\mathbf{d}_{\max} = o(\min\{n^{\frac{1}{4}}, \mathrm{Vol}(\mathbf{d}, v(H))^{\frac{1}{v(H)}}\})$. Thus we may assume that X is disconnected. We then apply the inductive hypothesis to each component of X to conclude the corollary. $\square$

4   Subgraph Counts

In this section, we determine the subgraph counts of a graph H in a random clustering graph. We will need the following lemma from [8] to prove Theorem 2.

Lemma 6. *Fix a* $\gamma \in [0,1]$, *and a vector* $\mathbf{d}$. *If* H *is a fixed graph and* ϕ *is an map from* $V(H)$ *to* $[n]$, *then the probability that* H *is a subgraph of* $G \hookleftarrow \mathcal{C}(\gamma, \mathbf{d})$ *at* ϕ *is the probability that at least one extension configuration* X *of* H *is a subgraph of* $G_0 \hookleftarrow \mathcal{G}(\mathbf{d})$ *for some map* $\psi : V(X) \to [n]$ *which restricts to* $\phi : V(H) \to [n]$ *and* X *induces* H, *i.e.,*

$$\mathop{\mathbb{P}}_{G \hookleftarrow \mathcal{C}(\gamma, \mathbf{d})} \left[H \curvearrowright_\phi G \right] = \mathop{\mathbb{E}}_{G_0 \hookleftarrow \mathcal{G}(\mathbf{d})} \left[\bigvee_{\substack{X \in \mathrm{Ext}(H) \\ \psi \in \mathrm{Hom}_\phi(X,[n])}} B_{\gamma, X} \cap X \curvearrowright_\psi G_0 \right] \quad (25)$$

where $B_{\gamma, X}$ *is the event that* X *produces a copy of* H *in the second stage of constructing* $G \hookleftarrow \mathcal{C}(\gamma, \mathbf{d})$.

Theorem 2. *Fix* $\gamma \in [0,1]$ *and a connected graph* H. *If* $\mathbf{d} \in \mathbb{R}^n$ *is a vector such that* $\delta_2(\mathbf{d})^{1+\Delta(H)} = o(\delta_3(\mathbf{d}))$ *and* $\mathbf{d}_{\max} = o(n^{\frac{1}{4}})$, *then the subgraph count of* H *in* $G \hookleftarrow \mathcal{C}(\gamma, \mathbf{d})$ *is determined by the* $v(H)$*th volume of* $\mathbf{d}$, *i.e.,*

$$\mathbb{E}_{G \hookleftarrow \mathcal{C}(\gamma, \mathbf{d})} \left[\mathrm{hom}(H, G) \right] = (1 + o(1)) \gamma^{e(H)} \mathrm{Vol}(\mathbf{d}, v(H)). \quad (26)$$

Proof. We first show the upper bound.

$$\mathop{\mathbb{E}}_{G \hookleftarrow \mathcal{C}(\gamma, \mathbf{d})} \left[\mathrm{hom}(H, G) \right] = \sum_{\phi : V(H) \to [n]} \mathop{\mathbb{P}}_{G_0 \hookleftarrow \mathcal{G}(\mathbf{d})} \left[H \curvearrowright_\phi G \right]$$

$$= \sum_{\phi : V(H) \to [n]} \mathop{\mathbb{P}}_{G_0 \hookleftarrow \mathcal{G}(\mathbf{d})} \left[\bigvee_{X \in \mathrm{Ext}(H)} \bigvee_{\psi \in \mathrm{Hom}_\phi(X,[n])} B_{\gamma, X} \cap X \curvearrowright_\psi G_0 \right] \quad (27)$$

$$\leq \sum_{\phi : V(H) \to [n]} \sum_{X \in \mathrm{Ext}(H)} \sum_{\psi \in \mathrm{Hom}_\phi(X,[n])} \mathop{\mathbb{P}}_{G_0 \hookleftarrow \mathcal{G}(\mathbf{d})} \left[B_{\gamma, X} \cap X \curvearrowright_\psi G_0 \right] \quad (28)$$

$$= \sum_{X \in \mathrm{Ext}(H)} \sum_{\psi \in \mathrm{Hom}(X,[n])} \mathop{\mathbb{P}}_{G_0 \hookleftarrow \mathcal{G}(\mathbf{d})} \left[B_{\gamma, X} \cap X \curvearrowright_\psi G_0 \right]$$

where we use Lemma 6 in Eq. (27) and the union bound in Eq. (28). Observe that the event $B_{\gamma, X}$ is independent of ψ. Thus,

$$\mathbb{E}_{G \hookleftarrow \mathcal{C}(\gamma, \mathbf{d})} \left[\mathrm{hom}(H, G) \right] = \sum_{X \in \mathrm{Ext}(H)} \mathbb{P}\left[B_{\gamma, X} \right] \sum_{\psi \in \mathrm{Hom}(X,[n])} \mathbb{P}_{G \hookleftarrow \mathcal{G}(\mathbf{d})} \left[X \curvearrowright_\psi G_0 \right]$$

$$= \sum_{X \in \mathrm{Ext}(H)} \mathbb{P}\left[B_{\gamma, X} \right] \mathbb{E}_{G \hookleftarrow \mathcal{G}(\mathbf{d})} \left[\mathrm{hom}(X, G_0) \right].$$

Applying Theorem 1 to each term in the sum, we find that

$$
\begin{aligned}
\mathbb{E}_{G \leftarrow \mathcal{C}(\gamma, \mathbf{d})}\left[\hom(H, G)\right] &\leq (1 + o(1))|\operatorname{Ext}(H)|)\, \mathbb{P}\left[B_{\gamma, K_{1,v(H)}}\right] \mathbb{E}_{G \leftarrow \mathcal{G}(\mathbf{d})}\left[\hom(K_{1,v(H)}, G_0)\right] \\
&= (1 + o(1))\, \mathbb{P}\left[B_{\gamma, K_{1,v(H)}}\right] \mathbb{E}_{G \leftarrow \mathcal{G}(\mathbf{d})}\left[\hom(K_{1,v(H)}, G_0)\right] \\
&= (1 + o(1)) \gamma^{e(H)} \operatorname{Vol}(\mathbf{d}, v(H))
\end{aligned}
$$

where we use the fact that there are a finite number of extension configurations and Lemma 1. Now we show a lower bound by the Bonferroni inequalities [3]. Again using Lemma 6, we have

$$
\begin{aligned}
\mathbb{E}_{G \leftarrow \mathcal{C}(\gamma, \mathbf{d})}\left[\hom(H, G)\right] &= \sum_{\phi: V(H) \to [n]} \mathbb{P}_{G_0 \leftarrow \mathcal{G}(\mathbf{d})}\left[\bigvee_{X \in \operatorname{Ext}(H)} \bigvee_{\psi \in \operatorname{Hom}_\phi(X, [n])} B_{\gamma, X} \cap X \curvearrowright_\psi G_0\right] \\
&\geq \sum_{\phi: V(H) \to [n]} \mathbb{P}_{G_0 \leftarrow \mathcal{G}(\mathbf{d})}\left[\bigvee_{\psi \in \operatorname{Hom}_\phi(K_{1,v(H)}, [n])} B_{\gamma, K_{1,v(H)}} \cap K_{1,v(H)} \curvearrowright_\psi G_0\right] \\
&\geq S_1 - S_2
\end{aligned}
$$

where

$$
S_1 = \sum_{\phi: V(H) \to [n]} \sum_{\psi \in \operatorname{Hom}_\phi(K_{1,v(H)}, [n])} \mathbb{P}_{G_0 \leftarrow \mathcal{G}(\mathbf{d})}\left[B_{\gamma, K_{1,v(H)}} \cap K_{1,v(H)} \curvearrowright_\psi G_0\right]
$$

$$
S_2 = \sum_{\phi: V(H) \to [n]} \sum_{\substack{\psi_1, \psi_2 \in \operatorname{Hom}_\phi(K_{1,v(H)}, [n]) \\ \psi_1 \neq \psi_2}} \mathbb{P}_{G_0 \leftarrow \mathcal{G}(\mathbf{d})}\left[B'_{\gamma, K_{1,v(H)}} \cap K_{1,v(H)} \curvearrowright_{\psi_1} G_0, K_{1,v(H)} \curvearrowright_{\psi_1} G_0\right]
$$

and $B'_{\gamma, K_{1,v(H)}}$ is the event that two distinct copies of $K_{1,v(H)}$ produce a copy of H.

By the same reasoning as in the upper bound, $S_1 = \gamma^{e(H)} \operatorname{Vol}(\mathbf{d}, v(H))$, so we focus on S_2. Observe that if $\psi_1 \neq \psi_2$ for $\psi_1, \psi_2 \in \operatorname{Hom}_\phi(K_{1,v(H)}, [n])$, the event that $K_{1,v(H)} \curvearrowright_{\psi_2} G_0$ and $K_{1,v(H)} \curvearrowright_{\psi_1} G_0$ is the same as the event that $K_{2,v(H)}$ appears as a subgraph of G_0 at the unique map ψ which restricts to ψ_1 and ψ_2, i.e., $K_{2,v(H)} \curvearrowright_\psi G_0$. Hence

$$
\begin{aligned}
S_2 &= \sum_{\phi: V(H) \to [n]} \sum_{\psi \in \operatorname{Hom}_\phi(K_{2,v(H)}, [n])} \mathbb{P}_{G \leftarrow \mathcal{G}(\mathbf{d})}\left[B'_{\gamma, K_{1,v(H)}} \cap K_{2,v(H)} \curvearrowright_\psi G_0\right] \\
&= \sum_{\psi \in \operatorname{Hom}(K_{2,v(H)}, [n])} \mathbb{P}_{G \leftarrow \mathcal{G}(\mathbf{d})}\left[B'_{\gamma, K_{1,v(H)}}, K_{2,v(H)} \curvearrowright_\psi G_0\right] \\
&= \mathbb{P}\left[B'_{\gamma, K_{1,v(H)}}\right] \sum_{\psi \in \operatorname{Hom}(K_{2,v(H)}, [n])} \mathbb{P}_{G \leftarrow \mathcal{G}(\mathbf{d})}\left[K_{2,v(H)} \curvearrowright_\psi G_0\right] \\
&= \mathbb{P}\left[B'_{\gamma, K_{1,v(H)}}\right] \mathbb{E}\left[\hom(K_{2,v(H)}, G_0)\right].
\end{aligned}
$$

By Lemma 1,

$$
\frac{\mathbb{E}_{G \hookleftarrow \mathcal{G}(\mathbf{d})} \left[\hom(K_{2,v(H)}, G_0) \right]}{\mathrm{Vol}(\mathbf{d}, v(H))} = \frac{\mathrm{Vol}(\mathbf{d}, 2)^{v(H)} \, \mathrm{Vol}(\mathbf{d}, v(H))}{\mathrm{Vol}(\mathbf{d})^{2v(H)}}
$$

$$
\leq \frac{\mathbf{d}_{\max}^{2v(H)-1}}{\mathrm{Vol}(\mathbf{d})^{v(H)-1}}
$$

$$
\leq n^{\frac{(2v(H)-1)}{4} - (v(H)-1)}
$$

$$
= o(1)
$$

where we use the fact that $v(H) > 1$ in the final line. It follows that $S_2 = o(S_1)$ and thus

$$
\mathbb{E}_{G \hookleftarrow \mathcal{C}(\gamma, \mathbf{d})} \left[\hom(H, G) \right] = (1 + o(1)) \gamma^{e(H)} \, \mathrm{Vol}(\mathbf{d}, v(H)) \tag{29}
$$

as desired.   □

Remark 4. By using Corollary 1 in place of Theorem 1, an analogous proof shows that if H has connected components $C_1, \ldots, C_p$, then

$$
\mathbb{E}_{G \hookleftarrow \mathcal{C}(\gamma, \mathbf{d})} \left[\hom(H, G) \right] = (1 - o(1)) \gamma^{e(H)} \prod_{i=1}^{p} \mathrm{Vol}(\mathbf{d}, |C_i|) \tag{30}
$$

Theorem 3. *Fix a connected graph H and $\gamma \in [0, 1]$. If $\mathbf{d} \in \mathbb{R}^n$ is a vector such that $\delta_2(\mathbf{d})^{2\Delta(H)+1} = o(\delta_3(\mathbf{d}))$ and $\mathbf{d}_{\max} \leq n^{\frac{1}{v(H)}}$, then $\hom(H, G)$ is concentrated around its mean with high probability over $G \hookleftarrow \mathcal{C}(\gamma, \mathbf{d})$.*

Proof. We aim to use Chebyshev's inequality. As

$$
\delta_2(\mathbf{d})^{\Delta(H)+1} \leq \delta_2(\mathbf{d})^{2\Delta(H)+1} o(\delta_3(\mathbf{d})) \tag{31}
$$

and $\mathbf{d}_{\max} \leq n^{\frac{1}{4}}$ we may apply Theorem 2 to find that, $\mathbb{E}_{G \hookleftarrow \mathcal{C}(\gamma, \mathbf{d})} \left[\hom(H, G) \right] = (1 + o(1)) \gamma^e(H) \, \mathrm{Vol}(\mathbf{d}, v(H))$. Let $\alpha(n)$ be a slowly-growing function to be specified later, and write $\overline{E} = (1 + o(1)) \gamma^e(H) \, \mathrm{Vol}(\mathbf{d}, v(H))$. By Chebyshev's inequality,

$$
\mathbb{P}_{G \hookleftarrow \mathcal{C}(\gamma, \mathbf{d})} \left[\left| \hom(H, G) - \overline{E} \right| > \frac{\overline{E}}{\alpha(n)} \right] = \mathbb{P}_{G \hookleftarrow \mathcal{C}(\gamma, \mathbf{d})} \left[\left| \hom(H, G) - \overline{E} \right| > \frac{\overline{E}}{\alpha(n)} \right]
$$

$$
\leq \alpha(n)^2 \left(\frac{\mathbb{E}_{G \hookleftarrow \mathcal{C}(\gamma, \mathbf{d})} \left[\hom(H, G)^2 \right]}{\overline{E}^2} - 1 \right)
$$

$$
= (1 + o(1)) \alpha(n)^2 \left(\frac{\mathbb{E}_{G \hookleftarrow \mathcal{C}(\gamma, \mathbf{d})} \left[\hom(H, G)^2 \right]}{\gamma^{2e(H)} \, \mathrm{Vol}(\mathbf{d}, v(H))^2} - 1 \right).
$$

We first write $\mathbb{E}_{G \hookleftarrow \mathcal{C}(\gamma, \mathbf{d})} \left[\hom(H, G)^2 \right]$ as a sum over subgraph counts in $\mathcal{G}(\mathbf{d})$ as follows.

$$
\mathbb{E}_{G \hookleftarrow \mathcal{C}(\gamma, \mathbf{d})} \left[\hom(H, G)^2 \right] = \sum_{\psi_1, \psi_2 : H \to [n]} \mathbb{P}_{G \hookleftarrow \mathcal{C}(\gamma, \mathbf{d})} \left[H \curvearrowright_{\psi_1} G, H \curvearrowright_{\psi_2} G \right].
$$

Let $\psi : V(H) \sqcup V(H) \to [n]$ be the map induced by a choice of ψ_1, ψ_2, and let U be the graph formed by identifying v_1 with v_2 in $H \sqcup H$ if $\psi_1(v_1) = \psi_2(v_2)$. We observe that the event that $H \curvearrowright_{\psi_1} G, H \curvearrowright_{\psi_2} G$ is the same as the event that $U \curvearrowright_\psi G$. There may be many pairs of maps ψ_1, ψ_2 which produce the same graph U, so we may reindex the above sum as

$$\mathbb{E}_{G \leftarrow \mathcal{C}(\gamma, \mathbf{d})} \left[\hom(H, G)^2 | \right] = \sum_{U \in Union(H, H)} \sum_{\psi : V(U) \to [n]} \mathbb{P}_{G \leftarrow \mathcal{C}(\gamma, \mathbf{d})} \left[U \curvearrowright_\psi G \right]$$

$$= \sum_{U \in Union(H, H)} \mathbb{E}_{G \leftarrow \mathcal{C}(\gamma, \mathbf{d})} \left[\hom(U, G) \right].$$

For $r \in \mathbb{N}$, let $Union_r(H, H)$ be the set of graphs in $Union(H, H)$ where the two copies of H share r vertices. For $r > 0$, $U \in Union_r(H, H)$ is connected since H is itself connected. Furthermore, $Union_0(H, H) = \{H \sqcup H\}$. By Theorem 2 and remark 4 we have

$$\mathbb{E}_{G \leftarrow \mathcal{C}(\gamma, \mathbf{d})} \left[\hom(H, G)^2 | \right] = (1 + o(1)) \left(\gamma^{2e(H)} \mathrm{Vol}(\mathbf{d}, v(H))^2 + \sum_{r > 0} \mathrm{Vol}(\mathbf{d}, 2v(H) - r) \left(\sum_{U \in Union_r(H, H)} \gamma^{e(U)} \right) \right).$$
(32)

For $r > 0$, we have

$$\frac{\mathrm{Vol}(\mathbf{d}, 2v(H) - r)}{\mathrm{Vol}(\mathbf{d}, v(H))^2} = \frac{\prod_{i=v(H)+1}^{2v(H)-r} \delta_i}{\mathrm{Vol}(\mathbf{d}, v(H))} \le \frac{\mathbf{d}_{\max}^{v(H)-r}}{\mathrm{Vol}(\mathbf{d}, v(H))} = o(1)$$
(33)

since $\mathbf{d}_{\max}^{v(H)} = o(\mathrm{Vol}(\mathbf{d}, v(H)))$. It follows that

$$\frac{\mathbb{E}\left[\hom(H, G)^2 \right]}{\gamma^{2e(H)} \mathrm{Vol}(\mathbf{d}, v(H))^2} = (1 + o(1)) + \sum_{r > 0} (1 + o(1)) | \frac{\mathrm{Vol}(\mathbf{d}, 2v(H) - r)}{\mathrm{Vol}(\mathbf{d}, v(H))^2} \left(\sum_{U \in Union_r(H, H)} \gamma^{e(U) - 2e(H)} \right)$$

$$= 1 + o(1).$$

We conclude

$$\mathbb{P}_{G \leftarrow \mathcal{C}(\gamma, \mathbf{d})} \left[\left| \hom(H, G) - \overline{E} \right| > \frac{\overline{E}}{\alpha(n)} \right] \le \alpha(n)^2 \left(1 + o(1) - 1 \right)$$
(34)

and the theorem follows by choosing $\alpha(n)$ growing sufficiently slowly so that the above expression is $o(1)$. □

5   Clustering Coefficient and Cycle Counts

In this section we apply Theorems 2 and 3 in several special cases. The *(global)* *clustering coefficient* of a graph G, denoted $C_0(G)$ is the ratio:

$$C_0(G) := \frac{\hom(K_3, G)}{\hom(P_2, G)}$$

where P_2 is the path with 2 edges.

Theorem 4. *Fix a vector* $\mathbf{d} \in \mathbb{R}^n$ *and* $\gamma \in [0,1]$. *If* $\delta_2(\mathbf{d})^5 = o(\delta_3(\mathbf{d}))$ *and* $\mathbf{d}_{\max} = o(n^{\frac{1}{4}})$, *then*

$$C_0(G) = (1 + o(1))\gamma \qquad (35)$$

with high probability over $G \hookleftarrow \mathcal{C}(\gamma, \mathbf{d})$.

Proof. Since $\Delta(K_3) = \Delta(P_2) = 2$, $\delta_2^{2\Delta(K_3)+1} = o(\delta_3)$. Furthermore, since $\mathbf{d}_{\max}^3 \leq n^{\frac{3}{4}} = o(n) \leq \mathrm{Vol}(\mathbf{d}, 3)$, we may apply Theorems 2 and 3 to conclude that with high probability

$$C_0(G) = \frac{(1 + o(1))\gamma^3 \, \mathrm{Vol}(\mathbf{d}, 3)}{(1 + o(1))\gamma^2 \, \mathrm{Vol}(\mathbf{d}, 3)} = (1 + o(1))\gamma.$$

□

Finally, we compute cycle counts.

Theorem 5. *Fix* $k \in \mathbb{N}$, *a vector* $\mathbf{d} \in \mathbb{R}^n$, *and* $\gamma \in [0,1]$. *If* $\delta_2(\mathbf{d})^3 = o(\delta_3(\mathbf{d}))$ *and* $\mathbf{d}_{\max} = o\left(\min\{n^{\frac{1}{4}}, \mathrm{Vol}(\mathbf{d}, k)\}\right)$, *then*

$$\mathbb{E}_{G \hookleftarrow \mathcal{C}(\gamma, \mathbf{d})} \left(\hom(C_k, G) \right) = (1 + o(1))\gamma^k \, \mathrm{Vol}(\mathbf{d}, k). \qquad (36)$$

If additionally $\delta_2(\mathbf{d})^5 = o(\delta_3(\mathbf{d}))$, *then*

$$\hom(C_k, G) = (1 + o(1))\gamma^k \, \mathrm{Vol}(\mathbf{d}, k). \qquad (37)$$

with high probability over $G \hookleftarrow \mathcal{C}(\gamma, \mathbf{d})$.

Proof. Since $\Delta(C_k) = 2$, $\delta_2(\mathbf{d})^{\Delta(H)+1} = o(\delta_3(\mathbf{d}))$. Since $\mathbf{d}_{\max}^k \leq o(\mathrm{Vol}(\mathbf{d}, k))$, we can apply Theorem 2 and the first claim follows. If $\delta_2(\mathbf{d})^5 = o(\delta_3(\mathbf{d}))$, then $\delta_2^{2\Delta(C_k)+1} = o(\delta_3)$ and thus we can apply Theorem 3 to conclude that with high probability

$$\hom(C_k, G) = (1 + o(1))\gamma^k \, \mathrm{Vol}(\mathbf{d}, k)$$

as desired. □

References

1. Aiello, W., Bonato, A., Cooper, C., Janssen, J., Prałat, P.: A spatial web graph model with local influence regions. Internet Math. **5**(1–2), 175–196 (2008). https://doi.org/10.1080/15427951.2008.10129305. http://www.internetmathematicsjournal.com/article/1458
2. Aiello, W., Chung, F., Lu, L.: Random evolution in massive graphs. In: Abello, J., Pardalos, P.M., Resende, M.G.C. (eds.) Handbook of Massive Data Sets. MC, vol. 4, pp. 97–122. Springer, Boston, MA (2002). https://doi.org/10.1007/978-1-4615-0005-6_4
3. Alon, N., Spencer, J.H.: The Probabilistic Method. John Wiley & Sons (2016)

4. Bloznelis, M., Karjalainen, J., Leskelä, L.: Normal and stable approximation to subgraph counts in superpositions of Bernoulli random graphs. J. Appl. Probab., 1–19 (2023). https://doi.org/10.1017/jpr.2023.48
5. Bloznelis, M., Leskelä, L.: Clustering and percolation on superpositions of Bernoulli random graphs. Random Struct. Algorithms **63**(2), 283–342 (2023). https://doi.org/10.1002/rsa.21140. https://onlinelibrary.wiley.com/doi/abs/10.1002/rsa.21140
6. Bradonjić, M., Hagberg, A., Percus, A.G.: The structure of geographical threshold graphs. Internet Math. **5**(1–2), 113–139 (2008)
7. Bringmann, K., Keusch, R., Lengler, J.: Geometric inhomogeneous random graphs. Theoret. Comput. Sci. **760**, 35–54 (2019)
8. Chung, F., Sieger, N.: A random graph model for clustering graphs. In: Dewar, M., Prałat, P., Szufel, P., Théberge, F., Wrzosek, M. (eds.) WAW 2023. LNCS, vol. 13894, pp. 112–126. Springer, Cham (2023). https://doi.org/10.1007/978-3-031-32296-9_8
9. Deijfen, M., Kets, W.: Random intersection graphs with tunable degree distribution and clustering (2015)
10. Krioukov, D., Papadopoulos, F., Kitsak, M., Vahdat, A., Boguná, M.: Hyperbolic geometry of complex networks. Phys. Rev. E **82**(3), 036106 (2010)
11. Young, S.J., Scheinerman, E.R.: Random dot product graph models for social networks. In: Bonato, A., Chung, F.R.K. (eds.) WAW 2007. LNCS, vol. 4863, pp. 138–149. Springer, Heidelberg (2007). https://doi.org/10.1007/978-3-540-77004-6_11

Self-similarity of Communities of the ABCD Model

Jordan Barrett[1], Bogumił Kamiński[2], Paweł Prałat[1(✉)],
and François Théberge[3]

[1] Department of Mathematics, Toronto Metropolitan University, Toronto, ON,
Canada
{jordan.barrett,pralat}@torontomu.ca
[2] Decision Analysis and Support Unit, SGH Warsaw School of Economics, Warsaw,
Poland
bkamins@sgh.waw.pl
[3] Tutte Institute for Mathematics and Computing, Ottawa, ON, Canada
theberge@ieee.org

Abstract. The Artificial Benchmark for Community Detection
(**ABCD**) graph is a random graph model with community structure and
power-law distribution for both degrees and community sizes. The model
generates graphs similar to the well-known **LFR** model but it is faster
and can be investigated analytically. In this paper, we show that the
ABCD model exhibits some interesting self-similar behaviour, namely,
the degree distribution of ground-truth communities is asymptotically
the same as the degree distribution of the whole graph (appropriately
normalized based on their sizes). As a result, we can not only estimate
the number of edges induced by each community but also the number of
self-loops and multi-edges generated during the process. Understanding
these quantities is important as (a) rewiring self-loops and multi-edges to
keep the graph simple is an expensive part of the algorithm, and (b) every
rewiring causes the underlying configuration models to deviate slightly
from uniform simple graphs on their corresponding degree sequences.

Keywords: Random graphs · Complex networks · Configuration
model · ABCD · Community structure · Self-similarity · Power-law

1 Introduction

One of the most important features of real-world networks is their community
structure, as it reveals the internal organization of nodes [7]. In social networks
communities may represent groups by interest, in citation networks they correspond to related papers, in the Web graph communities are formed by pages on
related topics, etc. Identifying communities in a network is therefore valuable as
this information helps us to better understand the network structure.

Unfortunately, there are very few datasets with ground-truth communities
identified and labelled. As a result, there is need for synthetic random graph
models with community structure that resemble real-world networks to benchmark and tune clustering algorithms that are unsupervised by nature. The **LFR**

© The Author(s), under exclusive license to Springer Nature Switzerland AG 2024
M. Dewar et al. (Eds.): WAW 2024, LNCS 14671, pp. 17–31, 2024.
https://doi.org/10.1007/978-3-031-59205-8_2

(**L**ancichinetti, **F**ortunato, **R**adicchi) model [19,20] is a highly popular model that generates networks with communities and, at the same time, allows for heterogeneity in the distributions of both node degrees and of community sizes. It became a standard and extensively used method for generating artificial networks.

A similar synthetic network to **LFR**, the **A**rtificial **B**enchmark for **C**ommunity **D**etection (**ABCD**) [14] was recently introduced and implemented[1], including a fast implementation[2] that uses multiple threads (**ABCDe**) [17]. Undirected variants of **LFR** and **ABCD** produce graphs with comparable properties but **ABCD/ABCDe** is faster than **LFR** and can be easily tuned to allow the user to make a smooth transition between the two extremes: pure (disjoint) communities and random graphs with no community structure. Moreover, it is easier to analyze theoretically—for example, in [13] various theoretical asymptotic properties of the **ABCD** model are investigated including the modularity function that, despite some known issues such as the "resolution limit" reported in [8], is an important graph property of networks in the context of community detection. Finally, the building blocks in the model are flexible and may be adjusted to satisfy different needs. Indeed, the original **ABCD** model was recently adjusted to include potential outliers (**ABCD+o**) [15] and extended to hypergraphs (**h-ABCD**) [16][3]. In the context of this paper, the most important of the above properties is that the **ABCD** model allows for theoretical investigation of its properties.

Another important aspect of complex networks is self-similarity and scale invariance which are well-known properties of certain geometric objects such as fractals [21]. Scale invariance in the context of complex networks is traditionally restricted to the scale-free property of the distribution of node degrees [1] but also applies to the distributions of community sizes [6,10], degree-degree distances [26], and network density [3]. Unfortunately, the definition of "scale free" has never reached a single agreement [5,11] but many experiments provide a statistical significance of these claims such as the experiment on 32 real-world networks that have a wide coverage of economic, biological, informational, social, and technological domains, with their sizes ranging from hundreds to tens of millions of nodes [26].

In search for more complete self-similar descriptions, methods related to the fractal dimension are considered that use box counting methods and renormalization [9,18,23]. However, the main issue is that complex networks are still not well defined in a proper geometric sense but one may, for example, introduce the concept of hidden metric spaces to overcome this problem [22].

For the context of community structure of complex networks, let us highlight one interesting study of the network of e-mails within a real organization that revealed the emergence of self-similar properties of communities [10]. Such

---

[1] https://github.com/bkamins/ABCDGraphGenerator.jl/.

[2] https://github.com/tolcz/ABCDeGraphGenerator.jl/.

[3] https://github.com/bkamins/ABCDHypergraphGenerator.jl.

experiments suggest that there is some universal mechanism that controls the formation and dynamics of complex networks.

In this paper, we show that the **ABCD** model exhibits self-similar behaviour: each ground-truth community inherits power-law degree distribution from the distribution of the entire graph (see Theorem 2), that is, the power-law exponent as well as the minimum degree of this distribution are preserved. On the other hand, as in all self-similarities mentioned above, some renormalization needs to be applied. In our case, the distribution is truncated so that the maximum degree, corrected by the noise parameter ξ (see Sect. 2 for its formal definition), does not exceed the community size.

The above observation, interesting and desired on its own, has some immediate implications that are of interest too. Firstly, we can easily compute the expected volume of each community (see Corollary 1). Secondly, and more importantly, we can investigate how many self-loops and multi-edges are constructed during the generation process of **ABCD** (see Theorem 3). Understanding this quantity is crucial for two reasons. Firstly, removing these self-loops and multi-edges to obtain a simple graph is a time consuming part of the construction algorithm. Secondly, as the **ABCD** construction involves several implementations of the well-known configuration model, the number of self-loops and multi-edges is directly correlated to how "skewed" the final graph is, i.e., more self-loops and multi-edges lead to distributions that are further away from being uniform. We speak about this second reason in more detail in Sect. 2.4.

The paper is structured as follows. In Sect. 2, we formally define the **ABCD** model and state one known result about the said model. The main results are presented in Sect. 3. Then, in Sect. 4, we present results of simulations that highlight properties that are proved in this paper and show their practical implications. The main result (Theorem 2) and its applications (Corollary 1 and Theorem 3) are proved in the long version of this paper. Finally, some open problems are presented in Sect. 5.

2   The ABCD Model

In this section we introduce the **ABCD** model. Its full definition, along with more detailed explanations of its parameters and features, can be found in [14]. We restate the main components of the **ABCD** model here to ensure completeness of the exposition in this article. More accurately, we outline a version of the **ABCD** model that was studied extensively in [13]. In the coming description, all choices made (the truncated power-law, the parameters, etc.) match those in [13]. In fact, there is much flexibility in the **ABCD** model, and we suspect that our results carry over to this more flexible setting. However, we choose to study the version of the **ABCD** model presented in [13] so that (a) we can use previously established results, and (b) we can simplify the statements of our main results.

2.1   Notation

For a given $n \in \mathbb{N} := \{1, 2, \ldots\}$, we use $[n]$ to denote the set consisting of the first n natural numbers, that is, $[n] := \{1, 2, \ldots, n\}$.

Our results are asymptotic by nature, that is, we will assume that $n \to \infty$. For a sequence of events $(E_n, n \in \mathbb{N})$, we say E_n holds *with high probability* (*w.h.p.*) if $\mathbb{P}(E_n) \to 1$ as $n \to \infty$. We say that E_n holds *with extreme probability* (*w.e.p.*) if $\mathbb{P}(E_n) = 1 - \exp(-\Omega(\log^2 n))$. In particular, if there are polynomially many events and each holds w.e.p., then w.e.p. all of them hold simultaneously.

Power-law distributions will be used to generate both the degree sequence and community sizes so let us formally define it. For given parameters $\gamma \in (0, \infty)$, $\delta, \Delta \in \mathbb{N}$ with $\delta \leq \Delta$, we define a truncated power-law distribution $\mathcal{P}(\gamma, \delta, \Delta)$ as follows. For $X \sim \mathcal{P}(\gamma, \delta, \Delta)$ and for $k \in \mathbb{N}$ with $\delta \leq k \leq \Delta$,

$$\mathbb{P}(X = k) = \frac{\int_k^{k+1} x^{-\gamma}\,dx}{\int_\delta^{\Delta+1} x^{-\gamma}\,dx}.$$

2.2   The Configuration Model

The well-known configuration model is an important ingredient of the **ABCD** generation process so let us formally define it here. Suppose then that our goal is to create a graph on n nodes with a given degree distribution $\mathbf{d} := (d_i, i \in [n])$, where $\mathbf{d}$ is a sequence of non-negative integers such that $m := \sum_{i \in [n]} d_i$ is even. We define a random multi-graph $\mathrm{CM}(\mathbf{d})$ with a given degree sequence known as the **configuration model** (sometimes called the **pairing model**), which was first introduced by Bollobás [4]. (See [2, 24, 25] for related models and results.)

We start by labelling nodes as $[n]$ and, for each $i \in [n]$, endowing node i with d_i half-edges. We then iteratively choose two unpaired half-edges uniformly at random (from the set of pairs of remaining half-edges) and pair them together to form an edge. We iterate until all half-edges have been paired. This process yields $G_n \sim \mathrm{CM}(\mathbf{d})$, where G_n is allowed self-loops and multi-edges and thus G_n is a multi-graph.

2.3   Parameters of the ABCD Model

The **ABCD** model is governed by the following eight parameters.

Parameter	Range	Description
n	$\mathbb{N}$	Number of nodes
γ	$(2, 3)$	Power-law degree distribution with exponentγ
δ	$\mathbb{N}$	Min degree as leastδ
ζ	$\left(0, \frac{1}{\gamma-1}\right]$	Max degree at mostn^ζ
β	$(1, 2)$	Power-law community size distribution with exponentβ
s	$\mathbb{N} \setminus [\delta]$	Min community size at least$s$
τ	$(\zeta, 1)$	Max community size at mostn^τ
ξ	$(0, 1)$	Level of noise

2.4   The ABCD Construction

We will use $\mathcal{A} = \mathcal{A}(n, \gamma, \delta, \zeta, \beta, s, \tau, \xi)$ for the distribution of graphs generated by the following 5-phase construction process.

Phase 1: Creating the Degree Distribution. In theory, the degree distribution for an **ABCD** graph can be any distribution that satisfies (a) a power-law with parameter γ, (b) a minimum value of at least δ, and (c) a maximum value of at most n^ζ. In practice, however, degrees are i.i.d. samples from the distribution $\mathcal{P}\left(\gamma, \delta, n^\zeta\right)$.

For $G_n \sim \mathcal{A}$, write $\mathbf{d}_n = (d_i, i \in [n])$ for the chosen degree sequence of G_n with $d_1 \geq \cdots \geq d_n$. Finally, to ensure that $\sum_{i \in [n]} d_i$ is even, we decrease d_1 by 1 if necessary; we relabel as needed to ensure that $d_1 \geq d_2 \geq \cdots \geq d_n$. This potential change has a negligible effect on the properties we investigate in this paper and we thus only present computations for the case when d_1 is unaltered.

Phase 2: Creating the Communities. We next assign communities to the **ABCD** model. When we construct a community, we assign a number of vertices to said community equal to its size. Initially, the communities will form an empty graph. Then, in Phases 3, 4 and 5, we handle the construction of edges using the degree sequence established in Phase 1.

Similar to the degree distribution, the distribution of community sizes must satisfy (a) a power-law with parameter β, (b) a minimum value of s, and (c) a maximum value of n^τ. In addition, we also require that the sum of community sizes is exactly n. Again, we use a more rigid distribution in practice: communities are generated with sizes determined independently by the distribution $\mathcal{P}(\beta, s, n^\tau)$. We generate communities until their collective size is at least n. If the sum of community sizes at this moment is $n + k$ with $k > 0$ then we perform one of two actions: if the last added community has size at least $k + s$, then we reduce its size by k. Otherwise (that is, if its size is $c < k + s$), then we delete this community, select c old communities and increase their sizes by 1. This again has a negligible effect on the analysis and we thus only present computations for the case when community sizes are unaltered.

For $G_n \sim \mathcal{A}$, write L for the (random) number of communities in G_n and write $\mathbf{C}_n = (C_j, j \in [L])$ for the chosen collection of communities in G_n with $|C_1| \geq \cdots \geq |C_L|$ (again, let us stress the fact that $\mathbf{C}_n$ is a random vector of random length L).

Phase 3: Assigning Degrees to Nodes. At this point in the construction of $G_n \sim \mathcal{A}$ we have a degree sequence $\mathbf{d}_n$ and a collection of communities $\mathbf{C}_n$ with community C_j containing $|C_j|$ *unassigned* nodes, i.e., nodes that have not been assigned a label or a degree. We then iteratively assign labels and degrees to nodes as follows. Starting with $i = 1$, let U_i be the collection of unassigned nodes at step i. At step i choose a node uniformly at random from the set of nodes u in U_i that satisfy

$$d_i \leq \frac{|C(u)| - 1}{1 - \xi\phi},$$

where $C(u)$ is the community containing u and

$$\phi = 1 - \frac{1}{n^2} \sum_{j \in [L]} |C_j|^2,$$

and assign this node label i and degree d_i; we have that $U_{i+1} = U_i \setminus \{u\}$. We bound the degrees assignable to node u in community C to ensure that there are enough nodes in $C \setminus \{u\}$ for u to pair with, preventing guaranteed self-loops or guaranteed multi-edges during phase 4 of the construction. The details of this bound are quite involved and are not overly important for our results. Thus, we point the reader to either [13] or [14] for a full explanation of the bound.

Phase 4: Creating Edges. At this point G_n contains n nodes labelled as $[n]$, partitioned by the communities $\mathbf{C}_n$, with node $i \in [n]$ containing d_i unpaired half-edges. The last step is to form the edges in G_n. Firstly, for each $i \in [n]$ we split the d_i half-edges of i into two distinct groups which we call *community* half-edges and *background* half-edges. For $a \in \mathbb{Z}$ and $b \in [0, 1)$ define the random variable $\lfloor a + b \rceil$ as

$$\lfloor a + b \rceil = \begin{cases} a & \text{with probability } 1 - b, \text{ and} \\ a + 1 & \text{with probability } b. \end{cases}$$

Now define $Y_i := \lfloor (1 - \xi)d_i \rceil$ and $Z_i := d_i - Y_i$ (note that Y_i and Z_i are random variables with $\mathbb{E}[Y_i] = (1 - \xi)d_i$ and $\mathbb{E}[Z_i] = \xi d_i$) and, for all $i \in [n]$, split the d_i half-edges of i into Y_i community half-edges and Z_i background half-edges. Next, for all $j \in [L]$, construct the *community graph* $G_{n,j}$ as per the configuration model on node set C_j and degree sequence $(Y_i, i \in C_j)$. Finally, construct the *background graph* $G_{n,0}$ as per the configuration model on node set $[n]$ and degree sequence $(Z_i, i \in [n])$. In the event that the sum of degrees in a community is odd, we pick a maximum degree node i in said community and replace Y_i with $Y_i + 1$ and Z_i with $Z_i - 1$. As we show in the proof of Theorem 3 in the long version of the paper, this minor adjustment also has a negligible effect on the analysis and we thus assume that none of these sums are odd. Note that $G_{n,j}$ is a graph and C_j is the set of nodes in this graph; we refer to C_j as a *community* and $G_{n,j}$ as a *community graph*. Note also that $G_n = \bigcup_{0 \leq j \leq n} G_{n,j}$.

Phase 5: Rewiring Self-loops and Multi-edges. Note that, although we are calling $G_{n,0}, G_{n,1}, \ldots, G_{n,L}$ graphs, they are in fact *multi-graphs* at the end of phase 4. To ensure that G_n is simple, we perform a series of *rewirings* in G_n. A rewiring takes two edges as input, splits them into four half-edges, and creates two new edges distinct from the input. We first rewire each community graph $G_{n,j}$ independently as follows.

1. For each edge $e \in E(G_{n,j})$ that is either a loop or contributes to a multi-edge, we add e to a *recycle* list that is assigned to $G_{n,j}$.
2. We shuffle the *recycle* list and, for each edge e in the list, we choose another edge e' uniformly from $E(G_{n,j}) \setminus \{e\}$ (not necessarily in the *recycle* list) and attempt to rewire these two edges. We save the result only if the rewiring does not lead to any further self-loops or multi-edges, otherwise we give up. In either case, we then move to the next edge in the *recycle* list.
3. After we attempt to rewire every edge in the *recycle* list, we check to see if the new *recycle* list is smaller. If yes, we repeat step 2 with the new list. If no, we give up and move all of the "bad" edges from the community graph to the background graph.

We then rewire the background graph $G_{n,0}$ in the same way as the community graphs, with the slight variation that we also add edge e to *recycle* if e forms a multi-edge with an edge in a community graph or, as mentioned previously, if e was moved to the background graph as a result of giving up during the rewiring phase of its community graph. At the end of phase 5, we have a simple graph $G_n \sim \mathcal{A}$.

Note that phase 5 of the **ABCD** construction process exists only to ensure that G_n is simple. Thus, if one were satisfied with a multi-graph G_n that had all of the properties $\mathcal{A}$ offers, one could simply terminate the process after phase 4. However, for most practical uses such as community detection, we require a simple graph and thus require phase 5. As mentioned in Sect. 1, phase 5 is a time consuming part of the algorithm. Theorem 3 gives us some insight as to why that is the case, namely, because with high probability the number of self-loops and multi-edges generated during phase 4 is at least $\Omega(L)$. Theorem 3 is therefore quite valuable as it lets us know when our choice of γ, β, ζ and τ will yield a best-case-scenario number of self-loops and multi-edges (in expectation).

Theorem 3 is also valuable for helping us understand how "skewed" the community graphs, along with the background graph, are with respect to graphs generated uniformly at random from the set of simple graphs on the respective degree sequences. In [12], Janson shows that if a graph is constructed as the configuration model on degree sequence **d**, followed by a series of rewirings, then a relatively small number of rewirings yields a distribution that is asymptotically equal (with respect to the total variation distance) to the uniform distribution on simple graphs with degree sequence **d**. By extrapolating this result, we can infer that the number of rewirings required in phase 5 of the **ABCD** construction process is directly correlated with how "skewed" the resulting graph is.

2.5   A Known Result for ABCD

A result from [13] that we use often in this paper is a tight bound on the number of communities generated by the **ABCD** model.

Theorem 1 ([13] Corollary 5.5 (a)). *Let $G_n \sim \mathcal{A}$ and let L be the number of communities in G_n. Then w.e.p. the number of communities, L, is equal to*

$$L = L(n) = \left(1 + O\left((\log n)^{-1}\right)\right) \hat{c} n^{1-\tau(2-\beta)},$$

where

$$\hat{c} = \frac{2 - \beta}{(\beta - 1)s^{\beta - 1}} .$$

Note that the concentration in Theorem 1 is a consequence of the bound $|C_j| \leq n^\tau$ for all communities C_j and fails if this bound is omitted.

3   Main Result

Our main result is a stochastic bound on the degree sequence of a given community in $\mathcal{A}$. For $G_n \sim \mathcal{A}$ with degree sequence $\mathbf{d}_n$, and for community graph $G_{n,j}$ with nodes from C_j, we make the following distinction: the *degree sequence of $G_{n,j}$* is the degree sequence of the community graph $G_{n,j}$, whereas the *degree sequence of C_j* is the subset of $\mathbf{d}_n$ containing the degrees of nodes in C_j. Hence, the degree sequence of C_j is $(d_v, v \in C_j)$ and the degree sequence of $G_{n,j}$ is $(Y_v, v \in C_j)$ where we recall that $Y_v = \lfloor (1 - \xi)d_v \rfloor$. The following two results, Theorem 2 and Corollary 1, are stated in terms of the degree sequences $(d_v, v \in C_j)$. However, both results can be easily restated in terms of the degree sequences $(Y_v, v \in C_j)$.

Theorem 2. *Let $G_n \sim \mathcal{A}$. Let C_j be a community in G_n with $|C_j| = z$ and let $\mathbf{c}_j$ be the degree sequence of community C_j. Next, let $\epsilon = \epsilon(n) = n^{-(\tau - \zeta)(2 - \beta)/2} = o(1)$, let*

$$\Delta_z = \min \left\{ \frac{z - 1}{1 - \xi\phi}, n^\zeta \right\},$$

and let X^- and X^+ be random variables with the following probability distribution functions on $\{\delta, \ldots, \Delta_z\}$:

$$\mathbb{P}\left(X^- = k\right) = \frac{\int_k^{k+1} x^{-\gamma} \, dx}{\int_\delta^{\Delta_z + 1} x^{-\gamma} \, dx}, \quad and$$

$$\mathbb{P}\left(X^+ = k\right) = \frac{\left(1 - \epsilon \mathbf{1}_{[k=\delta]}\right) \int_k^{k+1} x^{-\gamma} \, dx}{(1 - \epsilon) \int_\delta^{\delta + 1} x^{-\gamma} \, dx + \int_{\delta + 1}^{\Delta_z + 1} x^{-\gamma} \, dx} = (1 + o(1))\mathbb{P}\left(X^- = k\right).$$

Finally, let X be a uniformly random element of $\mathbf{c}_j$. Then w.h.p. X is stochastically bounded below by X^- and above by X^+.

The proof of Theorem 2 can be found in the long version of this paper. The power of this theorem is that it allows us to compare the structure of community graphs in $G_n \sim \mathcal{A}$ with the structure of graphs constructed via the configuration model on an i.i.d. degree sequence that is well understood. In this paper we provide two uses of this new and powerful tool. The first is a sharpening of Lemma 5.6 in [13], describing the volumes of communities in $G_n \sim \mathcal{A}$. For $X \sim \mathcal{P}(\gamma, \delta, \Delta)$, write

$$\mu_\ell(\gamma, \delta, \Delta) = \mathbb{E}\left[X^\ell\right], \tag{1}$$

and note in particular that $\mu_1(\gamma, \delta, n^\varsigma)$ is the expected degree of a node in $G_n \sim \mathcal{A}$. Next, for community C_j, define

$$\operatorname{vol}(C_j) := \sum_{v \in C_j} d_v \,.$$

Corollary 1. *Let $G_n \sim \mathcal{A}$, let C_j be a community in G_n with $|C_j| = z$, and let*

$$\Delta_z = \min \left\{ \frac{z-1}{1 - \xi \phi}, n^\varsigma \right\} \,.$$

Then, conditioned on the stochastic domination in Theorem 2,

$$\frac{\mathbb{E}\left[\operatorname{vol}(C_j)\right]}{z} = (1 + o(1))\, \mu_1(\gamma, \delta, \Delta_z) = \begin{cases} (1 + o(1))\, \mu_1(\gamma, \delta, n^\varsigma) & \text{if } z(n) \to \infty, \text{ and} \\ \Theta\left(\mu_1(\gamma, \delta, n^\varsigma)\right) & \text{otherwise.} \end{cases}$$

The second use of Theorem 2 that we present here is an analysis of the number of self-loops and multi-edges that are created during phase 4 of the construction process of $G_n \sim \mathcal{A}$. In practice, phase 5 of the **ABCD** construction can be computationally expensive. It is therefore valuable to study the number of collisions (self-loops and multi-edges) generated during phase 4 of the construction. The following theorem tells us that, although w.h.p. we can never do better than generating $\Omega(L)$ collisions, where L is the number of communities, we expect to see *at most* $O(L)$ collisions under certain restrictions on γ, β, ς, and τ.

Theorem 3. *Let $G_n \sim \mathcal{A}$ and define the following five variables depending on G_n.*

$S_c :=$ *The number of self-loops in community graphs after phase 4.*

$M_c :=$ *The number of multi-edge pairs in community graphs after phase 4.*

$S_b :=$ *The number of self-loops in the background graph after phase 4.*

$M_b :=$ *The number of multi-edge pairs in the background graph after phase 4.*

$M_{bc} :=$ *The number of background edges that are also community edges after phase 4.*

Then, conditioned on the stochastic domination in Theorem 2,

1. $\mathbb{E}\left[S_c\right] = O\left((n^{1-\tau(2-\beta)})(1 + n^{\varsigma(4-\gamma-\beta)})\right)$,
2. $\mathbb{E}\left[M_c\right] = O\left((n^{1-\tau(2-\beta)})(1 + n^{\varsigma(7-2\gamma-\beta)})\right)$,
3. $\mathbb{E}\left[S_b\right] = O(n^{\varsigma(3-\gamma)})$,
4. $\mathbb{E}\left[M_b\right] = O(n^{\varsigma(6-2\gamma)})$, *and*
5. $\mathbb{E}\left[M_{bc}\right] = o(\mathbb{E}\left[M_c\right])$.

Moreover, for all valid $\gamma, \beta, \varsigma, \tau$,

$$\mathbb{E}\left[S_c\right] = \Omega(L) \,,$$

if $\gamma + \beta > 4$ *then*

$$\mathbb{E}\left[S_c + M_c + M_{bc}\right] = \Theta(L),$$

if $2\zeta(3 - \gamma) + \tau(2 - \beta) \leq 1$ *then*

$$\mathbb{E}\left[S_b + M_b\right] = O(L),$$

and if both inequalities are satisfied then

$$\mathbb{E}\left[S_c + M_c + S_b + M_b + M_{bc}\right] = \Theta(L).$$

The proofs of Corollary 1 and Theorem 3, as well as the surrounding discussions, can be found in the long version of this paper.

4   Simulation Corner

In this section, we present a few experiments highlighting the properties that are proved to hold with high probability. The experiments show that the asymptotic predictions are useful even for graphs on a moderately small number of nodes.

4.1   The Coupling

Our main result (Theorem 2) shows that the degree distribution of a community of size z in $G_n \sim \mathcal{A}$ is stochastically sandwiched between $(X_i^-, i \in [z])$ and $(X_i^+, i \in [z])$ where $X_i^- \sim \mathcal{P}(\gamma, \delta, \Delta_z)$ and $X_i^+ \xrightarrow{d} X_i^-$ as $n \to \infty$. To compare the degree distribution of communities in **ABCD** graphs to the stochastic lower-bound $(X_i^-, i \in [z])$, we perform the following experiment. We generate three **ABCD** graphs G_n, G_n^* and G_n^{**}. Consistent in all three graphs are the parameters $n = 2^{20}, \delta = 5, \zeta = 0.4, s = 50, \tau = 0.6$, and $\xi = 0.5$. The graph G_n has parameters $\gamma = 2.1$ and $\beta = 1.1$, the graph G_n^* has $\gamma = 2.5$ and $\beta = 1.5$, and G_n^{**} has $\gamma = 2.9$ and $\beta = 1.9$. For each graph, we plot the complementary cumulative distribution function (ccdf) of degrees of (a) the whole graph, (b) the union of all smallest communities (G_n had 8 communities of size $s = 50$, G_n^* had 29, and G_n^{**} had 82), and (c) the unique largest community (sizes 4074, 4073, and 3903 in respective graphs G_n, G_n^*, and G_n^{**}). We then plot, in parallel, the expected ccdfs for the three graphs; for the whole graph the ccdf is that of $\mathcal{P}(\gamma, \delta, n^\zeta)$, and for the community graphs we use the expected ccdf of the stochastic lower-bound $(X_i^-, i \in [z])$, i.e., the function $\bar{F} : \{\delta, \ldots, \Delta_z\} \to [0, 1]$ where

$$\bar{F}(k) = \frac{\int_k^{\Delta_z + 1} x^{-\gamma}\, dx}{\int_\delta^{\Delta_z + 1} x^{-\gamma}\, dx} = \frac{k^{1-\gamma} - (\Delta_z + 1)^{1-\gamma}}{\delta^{1-\gamma} - (\Delta_z + 1)^{1-\gamma}}.$$

The results are presented in Fig. 1. From these results, we see that the distribution of $(X_i^-, i \in [z])$ is a very good approximation of the distribution of degrees in a community of smallest size as well as a community of largest size. We note that, since $(X_i^-, i \in [z])$ is a lower-bound, we expect the theoretical ccdf to sit slightly above the empirical ccdf, and this is confirmed by the experiment.

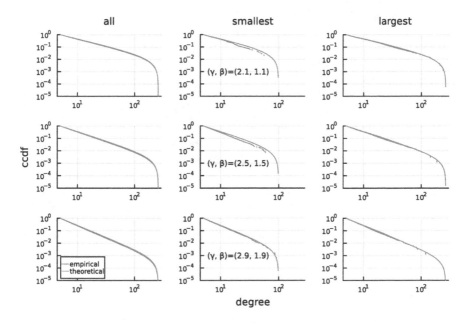

Fig. 1. The ccdf for the three different **ABCD** graphs G_n (top), G_n^* (middle), and G_n^{**} (bottom), and for three different subsets of nodes in each graph, namely, the whole graph (left), the union of smallest community graphs (middle), and the unique largest community graph (right). Each function is drawn on a log–log scale. The blue curves are the empirical data and the orange curves are the theoretical predictions. (Color figure online)

4.2   Volumes of Communities

Next, to investigate how well Corollary 1 predicts the volume of a particular community, we perform the following experiment. We generate three **ABCD** graphs G_n, G_n^* and G_n^{**}. Consistent in all three graphs are the parameters $n = 2^{20}$, $\delta = 5$, $\zeta = 0.6$, $s = 50$, $\tau = 0.9$, and $\xi = 0.5$. The graph G_n has parameters $\gamma = 2.1$ and $\beta = 1.1$, the graph G_n^* has $\gamma = 2.5$ and $\beta = 1.5$, and G_n^{**} has $\gamma = 2.9$ and $\beta = 1.9$. In each graph, we sorted communities with respect to their size (from the smallest to the largest) and then grouped them into 10 buckets as equal as possible (that is, the number of communities in any pair of buckets differs by at most one). For each bucket we compute the average degree and the standard deviation over all communities in that bucket. We compare it with the asymptotic prediction based on Corollary 1, that is, for each community of size z we compute $\mu_1(\gamma, \delta, \Delta_z)$, and take the average over all communities in the bucket. The results are presented in Fig. 2. We see that $n = 2^{20}$ is large enough and simulations match the theoretical predictions almost exactly.

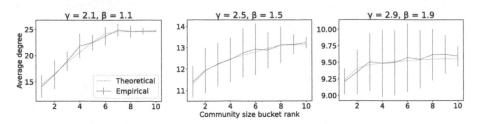

Fig. 2. The average degrees in communities for G_n (left), G_n^* (middle), and G_n^{**} (right). The communities are ranked by their size and grouped into 10 buckets as equal as possible. The blue line with error bars is the average degree and standard deviation among all communities in each bucket. Note that the errors, in absolute values, are largest for the leftmost plot and smallest for the rightmost plot. The orange dashed line shows the expected volumes for the stochastic lower-bound ($X_i^-, i \in [z]$), computed for each community size and bucketed in the same way as the empirical data. (Color figure online)

4.3   Self-loops and Multi-edges

Finally, to investigate the number of collisions (of various types) generated during phase 4 of the **ABCD** construction as functions of n, we perform the following experiment. For each $n \in \{2^{15}, 2^{16}, 2^{17}, 2^{18}, 2^{19}, 2^{20}\}$, we generate three sequences of 20 **ABCD** graphs $(G_n(i), i \in [20]), (G_n^*(i), i \in [20])$, and $(G_n^{**}(i), i \in [20])$. Consistent in all three sequences are the parameters $\delta = 5$, $\zeta = 0.6$, $s = 50$, $\tau = 0.9$, and $\xi = 0.5$. The graphs in sequence $(G_n(i), i \in [20])$ have $\gamma = 2.1$ and $\beta = 1.1$, the graphs in $(G_n^*(i), i \in [20])$ have $\gamma = 2.5$ and $\beta = 1.5$, and the graphs in $(G_n^{**}(i), i \in [20])$ have $\gamma = 2.9$ and $\beta = 1.9$. We compare the growth of S_c/L, M_c/L, S_b/L, and M_b/L (the average values and the corresponding standard deviations over 20 graphs), as functions of n, for all three sequences. Each sequence represents a different scenario in expectation based on Theorem 3, and we comment on each result separately.

- For $(G_n(i), i \in [20])$ with $\gamma = 2.1$ and $\beta = 1.1$, we have $\gamma + \beta < 4$ and $2\zeta(3 - \gamma) + \tau(2 - \beta) > \zeta(3 - \gamma) + \tau(2 - \beta) > 1$ and so we expect each of the variables S_c/L, M_c/L, S_b/L, and M_b/L to be unbounded. In Fig. 3 we see that, indeed, each of the four variables seem to grow with n in the simulations.
- For $(G_n^*(i), i \in [20])$ with $\gamma = 2.5$ and $\beta = 1.5$, we have $\gamma + \beta = 4$ and $2\zeta(3 - \gamma) + \tau(2 - \beta) > 1 > \zeta(3 - \gamma) + \tau(2 - \beta)$ and so we expect S_b/L to be bounded and S_c/L, M_c/L, M_b/L to be unbounded. As Fig. 4 shows, the simulations are consistent with the theory for S_c/L, M_c/L and S_b/L. However, the trend of M_b/L is unclear. Considering that $2\zeta(3 - \gamma) + \tau(2 - \beta) = 1.05$ in this case, it is reasonable that the growth of M_b/L should not reveal itself at this scale of n.
- For $(G_n^{**}(i), i \in [20])$ with $\gamma = 2.9$ and $\beta = 1.9$, we have $\gamma + \beta > 4$ and $1 > 2\zeta(3 - \gamma) + \tau(2 - \beta) > \zeta(3 - \gamma) + \tau(2 - \beta)$ and so we expect all of S_c/L, M_c/L, S_b/L, M_b/L to be bounded. Figure 5 again shows us that theory

matches simulations. We note the very slight upward trend of S_c/L and M_c/L, likely due to n being too small to see the asymptotic bound take hold.

We conclude that Theorem 3 does a good job at telling us the behaviour of S_c/L, M_c/L, S_b/L, and M_b/L for various γ and β, although the results are not as clear as the other experiments which would likely be resolved by taking larger values of n.

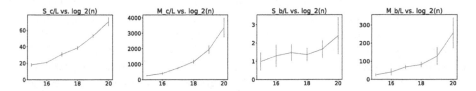

Fig. 3. In reading order: $S_c/L, M_c/L, S_b/L$ and M_b/L vs. $\log_2(n)$ for $(G_n(i), i \in [20])$, averaged over the 20 graphs.

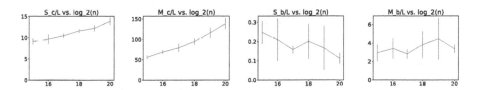

Fig. 4. In reading order: $S_c/L, M_c/L, S_b/L$ and M_b/L vs. $\log_2(n)$ for $(G_n^*(i), i \in [20])$, averaged over the 20 graphs.

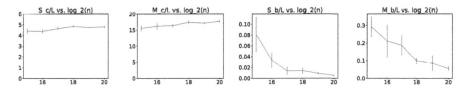

Fig. 5. In reading order: $S_c/L, M_c/L, S_b/L$ and M_b/L vs. $\log_2(n)$ for $(G_n^{**}(i), i \in [20])$, averaged over the 20 graphs.

5   Conclusion

Let us finish the paper with some open problems. We have shown two examples of how Theorem 2 can help us understand the nature of **ABCD** graphs. There are more applications of Theorem 2 that we do not explore here. Essentially, any result that holds for a configuration model on an i.i.d. degree sequence, sampled as $\mathcal{P}(\gamma, \delta, \Delta)$ for some $\gamma \in (2,3)$, should hold for a community graph in $G_n \sim \mathcal{A}$ modulo some discrepancy involving the rewiring phase of the **ABCD** construction. With additional work, it may also be true that such results hold for a community graph in $G_n \sim \mathcal{A}$. Possible avenues for $G_n \sim \mathcal{A}$ include studying its diameter, its diffusion rate, its clustering coefficient, etc.

Our results in Corollary 1 and Theorem 3 are results only in expectation, though our experiments indicate that the behaviour of at least community volumes is quite tight. Given that the truncated power-law $\mathcal{P}(\gamma, \delta, n^\varsigma)$ has unbounded second moment, and that $\mathcal{P}(\beta, s, n^\tau)$ has unbounded first moment, any study involving concentration will prove to be challenging. However, considering that the collection of community degree sequences partition the degree sequence of the whole graph, it is possible that these sequences exhibit self-correcting behaviour, and this is a potential road-map to a tighter version of our results.

In Theorem 3 we only show that collisions are bounded below asymptotically by $\Omega(L)$. On the other hand, our experimental results suggest that the number of collisions is, in fact, $\omega(L)$ when $\gamma + \beta \leq 4$ or when $2\zeta(3-\gamma) + \tau(2-\beta) > 1$. Thus, there is potential room to improve Theorem 3 by tightening the lower-bound.

References

1. Barabási, A.L., Albert, R.: Emergence of scaling in random networks. Science **286**(5439), 509–512 (1999)
2. Bender, E.A., Canfield, E.R.: The asymptotic number of labeled graphs with given degree sequences. J. Comb. Theory Ser. A **24**(3), 296–307 (1978)
3. Blagus, N., Šubelj, L., Bajec, M.: Self-similar scaling of density in complex real-world networks. Phys. A **391**(8), 2794–2802 (2012)
4. Bollobás, B.: A probabilistic proof of an asymptotic formula for the number of labelled regular graphs. Eur. J. Comb. **1**(4), 311–316 (1980)
5. Broido, A.D., Clauset, A.: Scale-free networks are rare. Nat. Commun. **10**(1), 1017 (2019)
6. Clauset, A., Newman, M.E., Moore, C.: Finding community structure in very large networks. Phys. Rev. E **70**(6), 066111 (2004)
7. Fortunato, S.: Community detection in graphs. Phys. Rep. **486**(3–5), 75–174 (2010)
8. Fortunato, S., Barthelemy, M.: Resolution limit in community detection. Proc. Nat. Acad. Sci. **104**(1), 36–41 (2007)
9. Gallos, L.K., Song, C., Makse, H.A.: A review of fractality and self-similarity in complex networks. Phys. A **386**(2), 686–691 (2007)
10. Guimera, R., Danon, L., Diaz-Guilera, A., Giralt, F., Arenas, A.: Self-similar community structure in a network of human interactions. Phys. Rev. E **68**(6), 065103 (2003)
11. Holme, P.: Rare and everywhere: perspectives on scale-free networks. Nat. Commun. **10**(1), 1016 (2019)
12. Janson, S.: Random graphs with given vertex degrees and switchings. Random Struct. Algorithms **57**(1), 3–31 (2020)

13. Kamiński, B., Pankratz, B., Prałat, P., Théberge, F.: Modularity of the ABCD random graph model with community structure. J. Complex Netw. **10**(6), cnac050 (2022)

14. Kamiński, B., Prałat, P., Théberge, F.: Artificial benchmark for community detection (ABCD)-fast random graph model with community structure. Netw. Sci., 1–26 (2021)

15. Kamiński, B., Prałat, P., Théberge, F.: Artificial benchmark for community detection with outliers (ABCD+o). Appl. Netw. Sci. **8**(1), 25 (2023)

16. Kamiński, B., Prałat, P., Théberge, F.: Hypergraph artificial benchmark for community detection (h–ABCD). J. Complex Netw. **11**(4), cnad028 (2023)

17. Kamiński, B., Olczak, T., Pankratz, B., Prałat, P., Théberge, F.: Properties and performance of the ABCDe random graph model with community structure. Big Data Res. **30**, 100348 (2022)

18. Kim, J., Goh, K.I., Salvi, G., Oh, E., Kahng, B., Kim, D.: Fractality in complex networks: critical and supercritical skeletons. Phys. Rev. E **75**(1), 016110 (2007)

19. Lancichinetti, A., Fortunato, S.: Benchmarks for testing community detection algorithms on directed and weighted graphs with overlapping communities. Phys. Rev. E **80**(1), 016118 (2009)

20. Lancichinetti, A., Fortunato, S., Radicchi, F.: Benchmark graphs for testing community detection algorithms. Phys. Rev. E **78**(4), 046110 (2008)

21. Mandelbrot, B.B.: The Fractal Geometry of Nature, vol. 1. WH Freeman New York (1982)

22. Serrano, M.Á., Krioukov, D., Boguná, M.: Self-similarity of complex networks and hidden metric spaces. Phys. Rev. Lett. **100**(7), 078701 (2008)

23. Song, C., Havlin, S., Makse, H.A.: Self-similarity of complex networks. Nature **433**(7024), 392–395 (2005)

24. Wormald, N.C.: Generating random regular graphs. J. Algorithms **5**(2), 247–280 (1984)

25. Wormald, N.C., et al.: Models of random regular graphs. In: London Mathematical Society Lecture Note Series, pp. 239–298 (1999)

26. Zhou, B., Meng, X., Stanley, H.E.: Power-law distribution of degree-degree distance: a better representation of the scale-free property of complex networks. Proc. Nat. Acad. Sci. **117**(26), 14812–14818 (2020)

A Simple Model of Influence: Details and Variants of Dynamics

Colin Cooper, Nan Kang, Tomasz Radzik$^{(\boxtimes)}$ (iD), and Ngoc Vu

Department of Informatics, King's College London, London, UK
tomasz.radzik@kcl.ac.uk

Abstract. We consider a simple model of establishing influence in a network. Vertices (people) split into influence groups and follow the opinion of the leader – the influencer – of their group. Groups can merge, based on interactions between influencers (the 'active vertices' of the network, while the followers are the 'passive vertices').

We study how the final number of influence groups depends on the way active vertices are chosen for interacting, considering two types of sparse graphs: the cycle C_n, which allows detailed analysis of various influencer algorithms, and random graphs $G(n, p)$ where $p = c/n$ for a constant c.

We also introduce a simple dynamic Falling-Out model, which allows for rejection of opinion. In its most general form, as considered for $G(n, p)$, one of the two interacting influencers can decide to follow the other influencer, or they both can reject the opinion of the other influencer and instead choose other influencers to follow.

Our analysis for the cycle is based on solving systems of recurrences using generating functions, and our analysis for the random $G(n, p)$ graph uses the differential equation method.

Keywords: Random graphs and processes · Social networks and influence

1 Introduction

We study a simple model of influence in a network. In the model vertices (people) follow the opinion of the group they belong to. This opinion percolates down from an active (or opinionated) vertex, the *influencer*, at the head of the group. Groups can merge, based on edges between influencers (active vertices), so that the number of opinions is reduced. Eventually no active edges (edges between influencers) remain and the groups and their opinions become static.

Interest in models of social structure was promoted by the sociologist Robert Axelrod [2], who posed the question "If people tend to become more alike in their

---

C. Cooper—Research by C. Cooper supported at the University of Hamburg, DFG Project 491453517.

M. Dewar et al. (Eds.): WAW 2024, LNCS 14671, pp. 32–46, 2024.
https://doi.org/10.1007/978-3-031-59205-8_3

beliefs, attitudes, and behavior when they interact, why do not all such differences eventually disappear?". This question was further studied by, for example, Flache *et al.* [8] and Moussaid *et al.* [12] who review various models of social interaction, generally based on some form of agency.

In our model, the emergence of separate groups occurs naturally due to lack of active edges between influencers. The exact composition of the groups and the opinion which influences them is a random outcome of the connectivity of the graph and the precise method of merging the groups. The paper [6] made an analysis of the Influencers model on an evolving version of the random graph $G(n, m)$. The graph has n vertices and m edges, which are selected randomly and added to the graph one at a time, in random order. If both end points of the current edge are active influencers, then one of them becomes a passive follower of the other, which remains as an active influencer. When no edges remain, the final influence structure is revealed (Fig. 1).

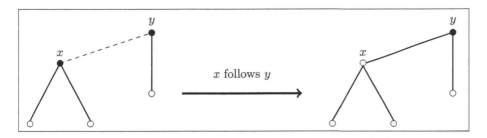

Fig. 1. Influencer x chooses a neighbouring influencer y and joins the group of y. Thus x becomes a follower of y along with the rest of its group.

In the current paper we study how the final number of influence groups depends on the way active vertices are chosen for interacting. We make this comparison for two types of sparse graphs: the cycle C_n, which allows detailed analysis of influencer algorithms; and random graphs $G(n, p)$ where each of the potential $\binom{n}{2}$ edges is present with probability $p = c/n$ for a constant c, independently of other edges.

We also introduce a simple dynamic model (Falling-Out) which allows for rejection of opinion. In its most general form, as considered for $G(n, p)$, one influencer can choose to follow another, or reject the opinion and instead follow another influencer. Although the dynamic is crude it does incorporate a natural component of human behaviour.

Joining Protocols. The Influencers problem is an instance of a general greedy process on a graph in which vertices are classified as active or passive. At any step, active vertices interact in some fashion, the result being that only one of the interacting vertices remains active, whilst the other interacting vertices become passive; the edges from the active vertex to the (now) passive ones being retained. This reduces the total number of active vertices until eventually there are no edges between them. At this point, the remaining active vertices become isolated and the process halts.

In this way, the process partitions the graph into disjoint subgraphs, which we call *fragments*, based on following the opinion of a neighbour. At any step, a fragment consists of a directed tree rooted at an active vertex, the current influencer, the edges pointing from follower vertices towards the root. This forms a simple model of influence where vertices in a fragment follow the opinion of the vertex they point to, and hence eventually that of the root.

The process is carried out on an underlying graph $G = (V, E)$. Initially all vertices of V are active, and all fragments are individual vertices. Let $A = A_t$ be the active set at step t and denote by $G[A]$ the subgraph of G induced by A. The simplest processes iterate the following steps.

1. *Vertex model*: Choose uniformly at random (u.a.r.) an active *non-isolated* vertex u and choose u.a.r. an active neighbour v of u. The edge uv is directed (u, v) and vertex u becomes passive (and follows the active vertex v).
2. *Edge model*: Choose a random edge from $G[A]$ and orient it u.a.r. If the oriented edge is (u, v), then u becomes passive (and follows the active vertex v).

Both processes end when $G[A]$ has no edges, and the final influencers are the final set A of remaining isolated active vertices.

Other processes studied here include variants of the edge model based on the active degrees of the edge endpoints and the following *Falling out (break edge) model*. Choose a random unexamined edge uv from $G[A]$. Vertices u and v then choose u.a.r. neighbours from $A \setminus \{u, v\}$, direct an edge to the chosen neighbours and become passive. If no such neighbour exists they become isolated.

Summary of Previous Results. The paper [6] made an analysis of the edge process on the random graph $G(n, m)$, choosing at step $1 \leqslant t \leqslant m \leqslant N \equiv \binom{n}{2}$ a random edge from the remaining $N - t + 1$ potential edges. Whenever both endpoints of the currently considered edge are active, one of them (selected u.a.r.) becomes passive. The results given in [6] include the following, among others.

- The number of fragments $a(m)$ in $G(n, m)$ is with high probability (w.h.p.) asymptotic to $F(m) = \frac{1}{1 - (1 - 1/n)\sqrt{1 - m/N}}$, for $m \ll N$, and this formula is an upper bound on the expected value of $a(m)$ for any $m \leqslant N(1 - o(1))$. The simulations indicate that this upper bound $F(m)$ may give the actual expected number of fragments for any $m \leqslant N(1 - o(1))$.
- The equivalent number of fragments in the presence of *stubborn vertices* (who accept followers, but refuse to follow). If $m = cN$, for a constant $c < 1$, then one stubborn vertex reduces the expected number of fragments to at most $\sqrt{1 - c}\, F(m)$.
- The sizes of the fragments correspond to the lengths of the parts of a stick in the *stick breaking process* [11,13]. The expected size of the largest of $k \geqslant 2$ remaining fragments is asymptotic to $(n/k) \log k$.

General Remarks. As previously noted, the Influencers problem is an instance of a class of greedy processes on graphs, in which at each step some subset of active vertices interact and all but one of them become passive. For the Influencer problem the interactions are pairwise based on the existence of edges between active vertices. The selected edges are retained to form tree components rooted at currently active vertices.

For a given graph G let $\alpha(G)$ be the independence number, the size of the largest independent set. The final set S of active isolated vertices returned by the Influencer process on a graph G is an independent set. Thus $|S| \leqslant \alpha(G)$, and it is natural to compare the size of S with the sizes of sets computed by heuristics for large independent sets.

The algorithm Greedy-IS (Greedy Independent Set) chooses at each step an active vertex v, adds v to S (initially empty), and deletes v and all of its neighbours from the graph, continuing until there are no vertices left in the graph. Greedy-IS fits formally the general description of the Influencers process, and can be viewed as a variant in which a selected active vertex v makes all its active neighbours passive (followers of v). Greedy-IS is a well known attempt to solve a *maximization* problem, whereas there is no obvious objective function for Influencers. Thus, realistically, protocols like the vertex and edge models align better with the underlying idea of modeling emergence of influencers than Greedy-IS, which requires that the selected active vertex imposes its opinion on *all* its neighbours.

The following Lemma, quoted from [5], is classic and gives a bound for Greedy-IS for any graph G.

Lemma 1. *Greedy-IS outputs an independent set S such that $|S| \geqslant n/(\Delta + 1)$ where Δ is maximum degree of G. This can be extended via Turan's theorem to prove $|S| \geqslant n/(d + 1)$ where d is the average degree.*

Since $\alpha(G)$, the independence number of G, is an upper bound on the final number of influencers, it may be tempting to assume Greedy-IS also gives an upper bound for the Influencer process. For $G(n, c/n)$ when $c > 1$ this should be the case, see [1], as Greedy-IS is notoriously difficult to improve. For very sparse graphs such as a cycle (or $G(n, c/n)$, $c < 1$) this is not necessarily so. In Sect. 2 we compare a range of algorithms for the Influencer process on C_n, one of which returns larger (independent) sets than Greedy-IS. In Sect. 3 we give a more limited comparison for $G(n, p)$. In both types of graph we also consider the Falling-Out dynamic model which has no direct relationship to independence number.

We define the *influencer ratio*, which is the limiting expected fraction of active isolated vertices S. Let G_n be a graph or graph space parameterized by n, for example C_n or $G(n, p)$. For a given fragmentation algorithm F, let S_n be the set of isolated active vertices remaining after running F on (a random element of) G_n and define the influencer ratio as

$$\rho(F) = \rho(F, G_n) = \lim_{n \to \infty} \mathbb{E}\,|S_n|/n.$$

The probabilities here are taken over the randomness in F and the potential randomness of G_n. Thus for the maximum independent set problem, $\rho_{IS} = \rho_{IS}(G_n) = \lim_{n \to \infty} \mathbb{E}\,\alpha(G_n)/n$.

2   The Influencer Problem on the Cycle C_n

As the independence number $\alpha(C_n) = \alpha(P_{n-1}) = \lfloor n/2 \rfloor$, we have $\rho(F) \leqslant \rho_{IS} = 1/2$ for any algorithm F applied to n-vertex cycle C_n or path P_n. The lower bound on the number of influencers is 1, achieved on P_n by the Left protocol, which keeps making the leftmost active vertex of the path passive; $\rho(\text{Left}) = 0$.

2.1   Results for the Cycle C_n

Greedy-IS (G-IS) was solved for the path P_n by Flory [9], who showed that the expected greedy independence ratio $\rho(\text{G-IS})$ tends to $\zeta_2 = (1 - e^{-2})/2$ as the path length tends to infinity. The cycle C_n is asymptotically equivalent to P_n.

The simplicity of the cycle allows us to compare many related strategies for the Influencer model. The algorithms listed in Table 1 are all randomized, and the stated value of $\rho(A)$ is the limiting expected value ratio. In [14] N. Vu studied many of these variants of the influencer problem on the cycle C_n, as an example of graph fragmentation. In most cases the value $\rho(A)$ is obtained via generating functions, as the solution to a system of recurrences. These results also hold asymptotically for any class of n-vertex 2-regular graphs with at most $o(n)$ cycles of finite length. In particular they hold w.h.p. for the underlying graph of a random permutation, as the expected number of cycles length at most ℓ in a random permutation is $O(\log \ell)$.

1. **Active vertex, active neighbour.** A random active vertex is chosen and a u.a.r. active neighbour (if any).
   (a) **Vertex Passive.** The chosen vertex becomes passive (or isolated).
   (b) **Neighbour Passive.** The chosen neighbour becomes passive.
   (c) **Lower Passive.** The lower degree vertex becomes passive. Ties are broken in favour of the neighbour.
   (d) **Higher Passive.** The higher degree vertex becomes passive. Ties are broken in favour of the chooser.
2. **Active vertex, maximum-degree active neighbour.** An active vertex is chosen and a random active neighbour of maximum degree.
   (a) **Overall Max. Deg. Wins.** The vertex of maximum degree remains active. Ties are broken randomly.
   (b) **Chooser Wins.** The choosing vertex remains active.
   (c) **Max. Deg. Neighbour.** The maximum degree neighbour wins. Ties are broken randomly.
3. **High-Low.** The lowest degree neighbour of a random vertex of highest degree is made passive. In the cycle, as long as paths of length at least 3 remain, a vertex of degree 2 is chosen. If it is next to a path endpoint, the endpoint becomes passive, and otherwise a random neighbour becomes passive.

4. **Random Edge.** A random edge between active vertices is chosen and one end becomes passive.
5. **Falling-Out (Dynamic).** A random edge between active vertices is broken and, each endpoint attaches to its random active neighbour, if any.

Table 1. Influencer algorithms on cycle C_n and their ratios.

Algorithm A	Influencer ratio $\rho(A)$	Value	Comments
Max. Indep. Set	$1/2$	0.5000	Upper bound
Greedy-IS	$\zeta_2 = (1 - e^{-2})/2$	0.4323	Flory [9]
Vertex Passive(VP)	$1/3$	0.3333	Sect. 2.2
Neighbour Passive	$1 - e^{-1/2}$	0.3935	
Lower Passive	$1 + e^{1/2} - \sqrt{2\pi}\,\mathrm{erfi}(1/\sqrt{2})$	0.2588	
Higher Passive	$1 - 1/2e - \sqrt{\pi/4}\,\mathrm{erf}(1)$	**0.4426**	Largest
Overall Max. Deg.	$\left[2(1 - e^{1/12}) - \int_0^1 (3x - 2)e^{-x^3/6+x^2/4}\right]/e^{1/12}$	0.2959	
Chooser	$\left[1 - e^{1/6} - \int_0^1 (x - 1)e^{-x^3/3+x^2/2}\right]/e^{1/6}$	0.2919	
Max. Deg. Nbr.	$1/3$	0.3333	Same as VP
High-Low	$\left[e^{5/6} - 1 - \int_0^1 xe^{-x^3/3+x^2/2}\right]/e^{5/6}$	**0.2366**	Smallest
Random Edge	$1/e$	0.3677	Sect. 2.2
Falling-Out	$\zeta_2 = (1 - e^{-2})/2$	0.4323	Dynamic

The largest value, $\rho \sim 0.44256$, is from Higher Passive (higher degree endpoint becomes passive). This supports the view that picking a vertex of minimum degree improves the performance of Greedy-IS. Similarly, the smallest value, $\rho \sim 0.2366$, is from High-Low see [14], where a minimum degree neighbour of a vertex of maximum degree is made passive.

For C_n, the dynamic variant Falling-Out (called Burn-Edge in [14]), has the same influencer ratio ζ_2 as the influencer ratio of Greedy-IS determined by Flory [9]. To see the correspondence between these two processes, use numbers $0, 1, \ldots, n - 1$ to label the consecutive vertices around the C'_n cycle for the Greedy-IS process and to label the consecutive edges around the C''_n cycle for the Falling-Out process. Couple the processes by selecting in the current step a random active vertex i in C'_n in the Greedy-IS process and using the edge with the same label i in C''_n in the Falling-Out process. To see that this coupling works, check by induction that at the beginning of each step a vertex i is active in C'_n in the Greedy-IS process, if and only if, the edge with the same label i is active in C''_n in the Falling-Out process. The processes end at the same step (no active vertices in C'_n and no active edges in C''_n). The number of vertices in

C_n' selected for the IS is the same as the number of broken edges in C_n'', which in turn is equal to the number of fragments in C_n'' computed by the Falling-Out process. (When the Falling-Out process terminates on C_n'', there is exactly one broken edge between each pair of consecutive fragments.)

2.2 Analysis for the Cycle C_n

In this section we give proofs for the Vertex Passive and Random Edge models. The Vertex Passive proof is short and simple and gives also the result for Maximum Degree Neighbour. The Random Edge proof is not given in [14]. For other cases the analysis is given in [14].

Vertex Passive algorithm (VP). The sampled active vertex follows its chosen active neighbour. Active vertices are sampled in a random order. This corresponds to a random permutation $(x_1, x_2, ..., x_n)$ of the vertex labels. When we process x_j, the ordered active set is $(x_{j+1}, ..., x_n)$. The vertices $x_1, ..., x_{j-1}$ are already either passive or isolated. To remove ambiguity, suppose vertex x_j follows the first occurrence of a neighbour (if any) in the sequence $(x_{j+1}, ..., x_n)$. If no such neighbour exists x_j is isolated.

Let y, z be the neighbours of a vertex x in C_n. The event that x is isolated occurs if y, z precede x in the random permutation. Of the six permutations of $\{x, y, z\}$, two put x last, so being isolated has probability $1/3$. Thus the expected number of active isolates is $n/3$.

Maximum Degree Neighbour (MDN). We explain why the expected number of fragments is the same as Vertex Passive (VP). On a cycle, VP and MDN act identically if the selected vertex has two active neighbours of the same degree, or only one active neighbour. Thus MDN can only differ from VP when a vertex next to a path endpoint is chosen, and the path is of length $L \geqslant 4$. Label the vertices $v_1, v_2, ..., v_L$. If vertex v_2 is chosen, in MDN it will attach to v_3 which has degree 2. The resulting active lengths of the sub-paths are 1 and $L - 2$, with v_1, v_3 active. In VP, choosing v_2 in a path of length at least 3, corresponds to a case where v_2 is first in the permutation, and thus has two active neighbours. Thus v_2 will attach to either v_1 or v_3 depending which is next in the permutation order. In either case, the resulting active lengths are 1 and $L - 2$.

Random Edge Algorithm. At each step a random edge between active vertices is chosen and one end becomes passive. An active path of length L is a (maximal) path of L active vertices, and thus $L - 1$ edges. A path of length one is an isolated active vertex. At step $t = 1$ a random edge of C_n is chosen and one endpoint becomes passive, resulting in an active path of length $n - 1$ bordered by the passive vertex. As the algorithm proceeds, this path is broken into smaller active paths bordered at each end by passive vertices, until finally only isolated vertices (active paths of length 1) remain.

Let N_L be the expected number of isolated vertices obtained by applying the Random Edge algorithm to an active path of length L. Thus $N_1 = 1$, $N_2 = 1$, and we define $N_0 = 0$. Given a path of length at least three, with vertices $v_1, ..., v_L$ and edges $e_1, ..., e_{L-1}$, where $e_i = (v_i, v_{i+1})$. Let $B(i, L)$ be the expected number of isolated vertices resulting when we apply the algorithm to edge e_i. Then

$$B(1, L) = B(L - 1, L) = \frac{1}{2}(N_1 + N_{L-2}) + \frac{1}{2}N_{L-1}$$

$$B(i, L) = \frac{1}{2}(N_{i-1} + N_{L-i}) + \frac{1}{2}(N_i + N_{L-i-1}), \quad 2 \leqslant i \leqslant L - 1.$$

It can be seen that $B(i, L) = B(L - i, L)$ for all i. Thus for $L \geqslant 3$,

$$N_L = \sum_{i=1}^{L-1} B(i, L) = \frac{1}{L-1}\left(N_1 + N_{L-2} + N_{L-1} + \sum_{i=2}^{L-2}(N_{i-1} + N_i)\right)$$

$$= \frac{1}{L-1}\sum_{i=1}^{L-1}(N_{i-1} + N_i), \quad \text{giving}$$

$$LN_L = N_L + \sum_{i=1}^{L-1}(N_{i-1} + N_i). \tag{1}$$

Observe that (1) holds also for $L = 2$ since $2N_2 = 2 = N_2 + (N_0 + N_1)$. Thus for $L \geqslant 3$, by subtracting from (1) the same formula for $L - 1$, we get

$$LN_L - (L - 1)N_{L-1} = N_L + N_{L-2}. \tag{2}$$

The expression (2) is also true for $L = 2$ and, by setting $N_{-1} = 0$, for $L = 1$. Let $G(x) = \sum_{L=1}^{\infty} N_L x^L$, multiply (2) by x^{L-1}, and sum up for $L \geqslant 1$ to obtain

$$G'(x) - xG'(x) = \frac{1}{x}G(x) + xG(x), \quad \text{which implies} \quad G'(x) = \frac{(x + 1/x)}{1 - x}G(x).$$

The general solution to this differential equation is

$$G(x) = A\frac{x}{(1 - x)^2}e^{-x}.$$

As $G(0) = 0$, differentiate again to obtain $G'(0) = A = N_1 = 1$. To find the coefficient $[x^n]G(x)$ write

$$G(x) = \frac{x}{(1 - x)^2}e^{-x} = \frac{x}{e(1 - x)^2}e^{1-x}$$

$$= \frac{x}{e(1 - x)^2}\left(1 + (1 - x) + \frac{(1 - x)^2}{2!} + \cdots + \frac{(1 - x)^j}{j!} + \cdots\right)$$

$$= \frac{xe^{-1}}{(1 - x)^2} + \frac{xe^{-1}}{(1 - x)} + f(x),$$

where $f(x)$ is entire (defined for all real numbers) so $\lim_{n\to\infty}[x^n]f(x) = 0$. Thus

$$[x^n]G(x) = ne^{-1} + e^{-1} + o_n(1),$$

and

$$\rho = \lim_{n\to\infty}\frac{N_{n-1}}{n} = e^{-1}.$$

3   The Influencer Problem for Random Graphs $G(n,p)$

The following is known about the independence number of $G(n,p)$; see [10, Theorem 7.4]. Let $\varepsilon > 0$ be a fixed constant, then for $c \geqslant c(\varepsilon)$, w.h.p.

$$\left|\alpha(G(n,c/n)) - \frac{2n}{c}\left(\log c - \log\log c - \log 2 + 1\right)\right| \leqslant \frac{\varepsilon n}{c}.$$

If $p = c/n$ where $c \to \infty$, then $\mathbb{E}\,\alpha \sim 2n\frac{\log c}{c}n$ and thus $\rho_{IS} \sim 2\frac{\log c}{c}$, whereas for c constant this is not the case.

3.1   Results for $G(n,p)$ When $p = c/n$

We give results for the Influencer model in $G(n,p)$ when $p = c/n$ for a constant c, for the following algorithms.

1. **Vertex Passive.** An active vertex is chosen u.a.r. and follows a random active neighbour.
2. **Random Edge.** Pick a random active edge and make one endpoint passive.
3. **Falling Out. (Dynamic)** A random edge between active vertices is chosen and broken. Each endpoint attaches to a random active neighbour, if any.

General Falling Out Model. We also consider the following generalisation of the Falling-Out model. With probability β, a random edge is broken and each endpoint vertex becomes passive by following another active neighbour, or isolated active, if no such neighbour exists. With probability $1 - \beta$ the edge is retained; in which case one endpoint becomes passive while the other remains active. If $\beta = 0$, the process is the Random Edge model; if $\beta = 1$, it is the basic Falling Out model. The general case of $0 \leqslant \beta \leqslant 1$ interpolates between these two models (Table 2).

Theorem 1. *In $G(n,p)$ with $p = c/n$ for a constant c, the Falling Out model with $0 \leqslant \beta \leqslant 1$ constant has the influencer ratio*

$$\rho_\beta = \frac{2(1 + \beta) - 2\beta e^{-c}}{2 + c(1 + \beta)}.$$

Table 2. Influencers ratios for algorithms on $G(n, p)$, where $p = c/n$ for an arbitrarily large constant c.

Algorithm A	Influencer ratio $\rho(A)$	$\lim_{c \to \infty} \rho(A)$	Comments
Maximum Indep. Set		$(2 \log c)/c$	For c large
Greedy-IS	$(\log(c + 1))/c$	$(\log c)/c$	
Vertex Passive	$(1 - e^{-c})/c$	$1/c$	
Random Edge	$2/(c + 2)$	$2/c$	
Falling-Out	$(2 - e^{-c})/(c + 1)$	$2/c$	Dynamic
Falling-Out(β)	$\dfrac{2(1 + \beta) - 2\beta e^{-c}}{2 + c(1 + \beta)}$	$2/c$	

3.2   Analysis for Random Graphs $G(n, p)$

Greedy-IS. For completeness, and as an introductory simple example of the methodology we use later, we sketch how the well-known ratio $\rho = \log(c + 1)/c$ of Greedy-IS on $G(n, c/n)$ can be derived. Let A_t be the number of active vertices at step t. The size of the computed independent set is $s = \min\{t : |A_t| = 0\}$. We have $A_0 = n$ and

$$\mathbb{E}\left(A_{t+1} \mid A_t\right) = \begin{cases} A_t - 1 - p(A_t - 1), & \text{if } A_t \geq 1, \\ 0, & \text{if } A_t = 0. \end{cases} \tag{3}$$

were $p(A_t - 1)$ is the expected number of neighbours of the selected vertex. Taking the expectation of both sides of (3) and rearranging, we get, where $q = 1 - p$,

$$\mathbb{E}\left(A_{t+1}\right) = q(\mathbb{E}\left(A_t\right) - 1) + q\,\mathbb{P}(A_t = 0).$$

This recurrence can be approximated (as long as $\mathbb{P}(A_t = 0) = o(1)$) with the recurrence $a_0 = n$, $a_{t+1} = q(a_t - 1)$, and one can show that if $0 \leq a_t = o(n)$, then $\mathbb{E}\left(A_t\right) = o(n)$ and $\mathbb{E}\left(s\right) = t + o(n)$. The recurrence for the sequence $(a_t)_{t \geq 0}$ solves to $a_t = (n + q/p)q^t - q/p$, giving $0 \leq a_t = o(n)$ for $t \sim \rho n$, $\rho = \log(c + 1)/c$.

Vertex Passive Model. At each step, random active vertex is chosen and follows a random active neighbour (if any). Thus the number of active vertices decreases by one. Initially there are n active vertices. So after t steps there are $n - t$ active vertices.

Let $S(t)$ be the number of isolated influencers at the start of step t. Thus $S(0) = 0$. As $S(t)$ increases by one if and only if the chosen vertex v has no active neighbours (no edges between v and any of the other $n - (t + 1)$ active neighbours), we have

$$\mathbb{E}\,S(t + 1) = S(t) + (1 - p)^{n-(t+1)},$$

$$\mathbb{E}\,S(n) = \sum_{t=0}^{n-1}(1 - p)^{n-t-1} = \frac{1 - (1 - p)^n}{1 - (1 - p)}.$$

Thus $\mathbb{E}\,S(n) \sim (1 - e^{-np})/p$, and if $p = c/n$, $\rho \sim (1 - e^{-c})/c$.

Other Models. In Sects. 3.3–3.5 we analyse the Random Edge and Falling-Out models. The analysis is related to the analysis of the random Greedy Matching presented in [7] and [10, Chapter 6.4]. Furthermore, the analysis is for the random $G(n, m)$ graph with $m = cn/2$ and we derive the influencer ratios ρ using the differential equation method. As any $m = cn/2 + o(n)$ gives the same ρ, and w.h.p. $G(n, p)$ for $p = c/n$ has $cn/2 + o(n)$ edges, the influencer ratios for $G(n, p = c/n)$ are the same as for $G(n, m = cn/2)$. We note that to obtain ρ, we are only interested in expected values and not w.h.p. results.

3.3   Random Edge

The following notation borrows from [10, Chapter 6.4] and [4] which give a proof of a related problem. Let $G(t) = (A_t, E_t)$ denote the active random subgraph of $G(n, m)$ remaining after t iterations. Let $\nu(t) = |A_t| = n - t$ be the number of vertices and $\mu(t)$ the number of edges in $G(t)$. At each step t, we choose a random edge $e_t = \{x, y\}$, delete the vertex x from A_t, and all edges incident with x. For the sake of our analysis we reveal the random graph as we run the algorithm.

A priori we do not know the degrees of any of the vertices. We just know that at step t we have $\nu(t)$ vertices and a uniform random set of $\mu(t)$ edges. We reveal the location of one of these edges, which is equally likely to have any two distinct endpoints among the $\nu(t)$ active vertices. More specifically, at each step t we reveal the edge e_t by choosing a random pair of distinct vertices. After we reveal the location of that edge, we know that its two endpoints must have degree at least 1. But we only know that because we revealed the edge.

For each edge e' among the other $\mu(t) - 1$ edges we reveal whether or not e' shares an endpoint with x. Any e' meeting x is deleted. Let $d'_t(x)$ be the number of such edges. Conditional on the edge e_t and the (say) $d'_t(x) = k$ deleted edges, the remaining edges comprise a uniform random set of $\mu(t) - 1 - k$ edges on the remaining set of $\nu(t) - 1$ vertices. Thus we have

$$\mathbb{E}\left[\mu(t+1) \mid \mu(t)\right] = \mu(t) - 1 - \mathbb{E}\left(d'_t(x) \mid \mu(t), e_t = \{x, y\}\right). \tag{4}$$

There are $1 \cdot (\nu(t) - 2)$ arrangements for an edge with one end at x and the other at a vertex other than x, y. Thus

$$\mathbb{E}\left(d'_t(x) \mid \mu(t), e_t = \{x, y\}\right) = (\mu(t) - 1) \cdot \frac{\nu(t) - 2}{\binom{\nu(t)}{2} - 1} = \frac{2(\mu(t) - 1)}{\nu(t) + 1}$$

$$= \frac{2\mu(t)}{\nu(t)} + O\left(\frac{1}{n - t} + \frac{\mu(t)}{(n - t)^2}\right). \tag{5}$$

Provided $t = dn$ for some constant $d < 1$, the error term on the RHS is $O(1/n)$.

From (4) and (5) we have

$$\mathbb{E}\left[\mu(t+1) \mid \mu(t)\right] = \mu(t) - 1 - \frac{2\mu(t)}{\nu(t)} + O\left(n^{-1}\right). \tag{6}$$

This leads us to consider the differential equation (DE) which will simulate the process w.h.p. Let $t = \tau n$, $M(\tau) = \mu(t)/n$, then

$$\frac{dM}{d\tau} = -1 - \frac{2M(\tau)}{1-\tau}, \qquad M(0) = \frac{c}{2}, \tag{7}$$

which has solution

$$M(\tau) = \frac{1}{2}(1-\tau)\,(c - \tau(c+2))\,.$$

The smallest positive root of $M(\tau) = 0$ is $\tau^* = c/(c+2)$, which gives

$$\rho = 1 - \tau^* = 2/(c+2). \tag{8}$$

It can be shown (see e.g. [4] or [10, Chapter 6.4]) that w.h.p. the process ends with an isolated active set of size $\tau^* n + o(n)$.

We only need expected values for the influencer ratio, and are using the DE method as a way to approximate the solution to non-standard recurrences; as we now explain. It can be checked that the solution $S(t) = nM(\tau)$ satisfies

$$S(t+1) = S(t) - 1 - \frac{2\,S(t)}{\nu(t)} + \frac{c+2}{n}.$$

Thus by the above and (6),

$$\mathbb{E}\,\mu(t+1) - S(t+1) = (\mathbb{E}\,\mu(t) - S(t))(1 - 1/(n-t)) + d_t/n.$$

Provided $t = \tau n$ for some constant $\tau < 1$, iterating this back to $S(0) = \mu(0) = cn/2$ the error term on the RHS is $O(t/n)$. So $\mathbb{E}\,\mu(t^*) = S(t^*) + O(1)$, and $\mathbb{E}\,\mu(t) = 0$ at some $t \sim t^*$.

3.4   Basic Falling-Out Model

We continue using the ideas in the formulation above, giving a brief proof, and leaving aside details. When the random edge $e_t = \{x, y\}$ is exposed, then given $\mu(t)$ and $\nu(t)$, the random graph $G(t)$ is otherwise unknown. We delete both x and y, the edge $\{x, y\}$ and the $d'_t(x) + d'_t(y)$ remaining edges adjacent to x or y. Thus we have $\nu(t) = n - 2t$ and the analysis of remaining edges follows the random Greedy Matching algorithm, see [7] or [10, Chapter 6.4]. The recurrence

$$\mathbb{E}\,(\mu(t+1) \mid \mu(t)) = \mu(t) - 1 - \mathbb{E}\,(d'_t(x) + d'_t(y) \mid \mu(t), e_t = \{x, y\})$$

gives rise to the following differential equation, analogous to (7),

$$\frac{dM}{d\tau} = -1 - \frac{4M(\tau)}{1 - 2\tau}, \qquad M(0) = \frac{c}{2}.$$

As before we used $\mathbb{E}\,d_t(x) = 1 + 2\mu/\nu + O(1/n)$, see (5). Hence

$$M(\tau) = \frac{1}{2}(1 - 2\tau)\,(c - 2(c+1)\tau)\,. \tag{9}$$

Thus $M(\tau) = 0$ at $\tau^* = c/2(c+1)$; the expected proportional size of the final matching (the set of removed independent edges) is $c/2(c+1)$ (see [7]) and an expected proportion of $1 - 2\tau^* = 1/(c+1)$ remaining isolated vertices.

It remains to calculate the number of isolated vertices created when edges were being deleted. An isolated vertex was created whenever an endpoint of the selected edge did not have any other neighbours. Let $d'_t(x)$ be the other edges incident with x after edge $\{x, y\}$ was exposed. Using a balls-in-boxes model where we throw $2(\mu(t) - 1)$ edge endpoints into $\nu(t)$ boxes, the probability none of the remaining randomly allocated edge endpoints is incident with x is

$$\mathbb{P}(d'_t(x) = 0) = \left(1 - \frac{1}{\nu(t)}\right)^{2(\mu(t)-1)} \sim e^{-2\mu/\nu}.$$

Conditional on this a similar result holds for y.

Let $S(t)$ be the number isolated active vertices arising from edge deletion at step t, then given $\mu(t)$ and $\nu(t) = n - 2t$,

$$\mathbb{E}\, S(t+1) = S(t) + 2(1 + o(1))e^{-2\mu(t)/n-2t}, \qquad S(0) = 0.$$

Let $t = t/n$, $M(\tau) = \mu/n$, $\sigma(\tau) = S(t)/n$, then using (9),

$$\frac{d\sigma}{d\tau} = 2e^{-M(\tau)/(1-2\tau)} = 2e^{-c}e^{2(c+1)\tau}, \qquad \sigma(0) = 0.$$

This has solution

$$\sigma(t) = \frac{e^{-c}}{c+1}\left(e^{2(c+1)\tau} - 1\right).$$

At $\tau^* = c/2(c+1)$, $\sigma(\tau^*) = (1 - e^{-c})/(c+1)$. Combining this with the $1 - 2\tau^* = 1/(c+1)$ remaining fraction of isolated vertices, we obtain the influencers ratio ρ for the basic Falling-Out model as

$$\rho = \frac{2 - e^{-c}}{c+1}. \tag{10}$$

3.5   General Falling-Out Model

For β constant, $0 \leqslant \beta \leqslant 1$, $\mathbb{E}\,\nu(t) = n - (1+\beta)t$. It follows as above that

$$\mathbb{E}\,\mu(t+1) \sim \mu(t) - 1 - \left(\frac{2\mu}{\nu} + \beta\frac{2\mu}{\nu}\right).$$

This gives the corresponding differential equation and its solution:

$$\frac{dM(\tau)}{d\tau} = -1 - \frac{2(1+\beta)M}{1 - (1+\beta)\tau}, \qquad M(0) = c/2,$$

$$M(\tau) = \frac{1}{2}(1 - (1+\beta)\tau)(c - \tau(2 + c(1+\beta))).$$

Thus $M(\tau) = 0$ at

$$\tau^* = \frac{c}{2 + c(1+\beta)}. \tag{11}$$

The expected number of isolated (active) vertices arising from edge deletion is

$$\mathbb{E}\, S(t+1) \sim S(t) + 2\beta e^{-2\mu/\nu}.$$

Putting $\sigma(\tau) = S(t)/n$, $\tau = t/n$, and $2\mu/\nu = c - \tau(2 + c(1+\beta))$ gives

$$\frac{d\sigma(\tau)}{d\tau} = 2\beta e^{-c} e^{\tau(2+c(1+\beta))}, \qquad \sigma(0) = 0,$$

which has solution

$$\sigma(\tau) = \frac{2\beta e^{-c}}{2 + c(1+\beta)} \left(e^{\tau(2+c(1+\beta))} - 1 \right).$$

It follows that

$$\sigma(\tau^*) = \frac{2\beta}{2 + c(1+\beta)} (1 - e^{-c}).$$

The expected fraction of isolated active vertices remaining at this point is $1 - (1+\beta)\tau^*$. Adding this to the above gives

$$\rho = \frac{2(1+\beta) - 2\beta e^{-c}}{2 + c(1+\beta)}.$$

In particular, $\beta = 0$ gives ρ for the Random Edge model as in (8), and $\beta = 1$ gives ρ for the Basic falling-Out model as in (10).

3.6 Formalizing the DE for w.h.p. Results

The formulations above use the Differential Equation method as a device to estimated expected values; the method to do this being shown at the end of Sect. 3.3.

Our analysis uses the notation and methodology of [10, Chapter 6.4], which gives an exposition of random Greedy Matching in $G(n,m)$ for $m = cn$, c constant. (Note that our analysis is for $m = cn/2$.) Thus although we do not require it, our results hold w.h.p. and not just in expectation. The models we analysed are similar to Greedy Matching. We refer the reader to [10, Chapter 6.4] for details, noting that the conditions (P1)-(P4) and the definition of the domain D as given there differ by constants from our setting, while the event $\mathcal{E}$ that the maximum degree $\Delta(G) \leqslant \log n$ is identical. See also [3,15] for further details.

4   Conclusions and Further Work

The simplicity of the cycle allows us to analyse a range of related algorithms for the Influencer process, a number of which we were able to generalize to $G(n,p)$ or

$G(n, m)$. It would be interesting to analyse more realistic variants of the Falling-Out model. For example an active vertex x picks an active neighbour y_1 and either follows y_1 or breaks the edge. In the latter case x picks another active neighbour y_2 and repeats this behaviour until they either find some neighbour y_k they agree with or become an isolated influencer. This may possibly be a more realistic model of social behaviour.

References

1. Coja-Oghlan, A., Efthymiou, C.: On independent sets in random graphs. Random Struct. Algorithms **47**(3), 436–486 (2015)
2. Axelrod, R.: The dissemination of culture: a model with local convergence and global polarization. J. Conflict Resolut. **41**(2), 203–226 (1997)
3. Bennett, P., Dudek, A.: A gentle introduction to the differential equation method and dynamic concentration. Disc. Math. **345**(12), 113071 (2022)
4. Bennett, P., Cooper, C., Frieze, A.: Rainbow Greedy Matching Algorithms. https://arxiv.org/abs/2307.00657 (2023)
5. Chekuri, C.: University of Illinois Urbana-Champaign. Course notes (2016). https://courses.engr.illinois.edu/cs583/sp2018/Notes/packing.pdf
6. Cooper, C., Kang, N., Radzik, T.: A simple model of influence. In: WAW 2023: Algorithms and Models for the Web Graph, pp. 164–178 (2023)
7. Dyer, M., Frieze, A.M., Pittel, B.: The average performance of the greedy matching algorithm. Ann. Appl. Probab. **3**, 526–552 (1993)
8. Flache, A., et al.: Models of social influence: towards the next frontiers. JASSS **20**(4), 2 (2017). http://jasss.soc.surrey.ac.uk/20/4/2.html
9. Flory, P.J.: Intramolecular reaction between neighboring substituents of vinyl polymers. J. Am. Chem. Soc. **61**(6), 1518–1521 (1939)
10. Frieze, A.M., Karoński, M.: Introduction to Random Graphs. CUP (2016). https://www.math.cmu.edu/~af1p/Book.html
11. Holst, L.: On the lengths of the pieces of a stick broken at random. J. Appl. Prob. **17**, 623–634 (1980)
12. Moussaïd, M., Kämmer, J.E., Analytis, P.P., Neth, H.: Social influence and the collective dynamics of opinion formation. PLoS ONE **8**(11), e78433 (2013)
13. Pyke, R.: Spacings. JRSS(B) **27**(3), 395–449 (1965)
14. Vu, N.: Some approaches to graph fragmentation with application to clustering geo-tagged data. Ph.D. thesis, King's College London (2018). https://kclpure.kcl.ac.uk/portal/en/persons/ngoc-vu/studentTheses/
15. Wormald, N.: The differential equation method for random graph processes and greedy algorithms. In: Lectures on Approximation and Randomized Algorithms, PWN, Warsaw, pp. 73–155 (1999)

Impact of Market Design and Trading Network Structure on Market Efficiency

Nick Arnosti[1] , Bogumił Kamiński[2] , Paweł Prałat[3] ,
and Mateusz Zawisza[2]([⊠])

[1] Department of Industrial and Systems Engineering, University of Minnesota,
207 Church Street SE, Minneapolis, MN 55455, USA
arnosti@umn.edu
[2] Decision Analysis and Support Unit, SGH Warsaw School of Economics,
Niepodległości 162, 02-554 Warsaw, Poland
{bogumil.kaminski,mzawisz}@sgh.waw.pl
[3] Department of Mathematics, Toronto Metropolitan University, 350 Victoria Street,
Toronto, ON, Canada
pralat@torontomu.ca

Abstract. This paper investigates the influence of market design, market size, and trading network structure on market efficiency and trade participation rate. The study considers two market designs: Zero Intelligence Traders (ZIT) in Chamberlin's bilateral haggling market and a greedy matching of traders on a network. Sellers and buyers are embedded in a random bipartite graph with varying network densities, and markets vary in size from 20 to 2000 traders.

Simulations reveal that greedy matching generally leads to more efficient allocations than ZIT trading networks. By increasing the average degree of a trading network from 1 to 5 or 10, market efficiency can be significantly improved for both market designs, achieving 89% and 95% of maximum efficiency, respectively. The study also contradicts the common belief that larger markets are better, as no significant impact of market size was found. We discuss the policy implications of these results.

Keywords: Zero-Intelligence Traders · bilateral exchange · market efficiency · Hungarian Algorithm · greedy matching algorithm · network science · agent-based simulation (ABM) · bipartite graph

1 Research Objective and Paper Structure

The primary objective of this research paper is to evaluate the collective impact of market design, market size, and trading network structure on market efficiency and trade participation rate. The research problems are identified in Sect. 2. The paper embarks on a comparative analysis of two distinct market designs described in detail in Sect. 3: the Zero Intelligence Traders (ZIT) model in Chamberlin's bilateral haggling market defined and characterized in Sect. 3.1

© The Author(s), under exclusive license to Springer Nature Switzerland AG 2024
M. Dewar et al. (Eds.): WAW 2024, LNCS 14671, pp. 47–64, 2024.
https://doi.org/10.1007/978-3-031-59205-8_4

and a model based on the greedy matching of traders on a random bipartite graph discussed in Sect. 3.2. This preliminary investigation aims to shed light on complex interplay between all three above factors of market design, market size and trading network structure, and their implications for market performance and trader engagement presented respectively in Sects. 4.1 and 4.2. Recommendations for policy makers can be found in Sect. 4.1. The final conclusions, along with the limitations of the research and directions for future research, are discussed in Sect. 5.

2   Definition of Research Problem and Its Motivation

The objective of this section is to delineate the research gaps and formulate pertinent questions, derived from the literature review presented in this section and the subsequent Sect. 3. These questions will be partially addressed in Sect. 4.

Studies on bilateral trading, typically employing experimental or simulation methods, have demonstrated a varied degree of market efficiency dependent on market designs. These range from approximately 70% of market efficiency for Chamberlin's Bilateral Haggling Market, as shown in the experimental studies of [2,3], to 90% or more of market efficiency for the double auction market mechanism. This high efficiency was demonstrated experimentally in [20,21] and through simulation methods in [8]. Both mechanisms operate under the assumption that traders can interact with everyone else. Although in practice, traders may only interact with a few other traders, the theoretical possibility of universal interaction exists. In real-world scenarios, we typically do not have access to everyone and must limit ourselves to a group of friends or acquainted traders. This observation underscores the importance of social networks in trading, as discussed in [6,10].

Given the observation that most of social interactions happen between a limited group of friends or acquainted traders, see [9,10], suggests that a natural extension of models proposed in the literature would be to embed traders in a social network and assess how this feature of real-world phenomena would impact key metrics of such markets, i.e. market efficiency and trade participation. To achieve this objective, authors of this contribution propose the following research agenda of traders on networks that is depicted in Fig. 1.

Figure 1 summaries what has been already established in the literature and what is still missing, i.e. constitutes a research gap, and what potential direction of further research can be. This agenda is intended to investigate two drivers of market efficiency:

- market design, i.e. how specific market mechanisms, e.g. Chamberlin's higgling market [2,3], greedy matching, Hungarian algorithm [5,6,19], perfect competition [16,22], impact the market efficiency and trade participation,
- social network structure, i.e. how network structure, e.g. its density, size, clustering [6,10,17], impact key market outcomes.

Researched thoroughly Partially researched (gap) Not researched (gap)	Bi-partie graph stuctures of buyers and sellers		
	Complete bipartite graph, i.e. everybody can trade with everybody	Δ = Impact of network on efficiency	Sparse bi-partie graphs of buyers and sellers
Intelligence Social planner maximizing social wefare, i.e. consumer surplus + profits	**Classical equilibrium price and volume** in demand & supply model, i.e. matching only so called intramarginal buyers and seller, instead of extramarg.	Impact of network for intelligent market design	**Optimal market-clearing prices** solution for graphs, see Damage, et al. (1986), aka **Hungarian Algorithm** → lack of analytical results for efficiency loss
Δ = Impact of market design and **intelligence**	Impact of market design for full graph	Combined Impact	Impact of market design for sparse graph
Zero-intelligence Traders (Chamberlin's Bilateral Haggling ZIT market, DA), i.e. bottom-up self-organizaing market design	• Simulation and experimental results for **Double Auction**, see Gode, Sunder (1993) • Only experimental, but no simulation or analytical results for **Chamberlin's ZIT**	Impact of network	**Obtaining analytical results** for tractable bi-partie graph structure or demand & supply functions supplemented by simulation results for less tractable cases

(left margin label, vertical: **Market design**)

Fig. 1. Potential research agenda of traders on networks **Source:** Own work

Two directions of intervention can be investigated separately or as a whole, but still one would like to decompose the final effect into the effect of market design and the effect of network structure.

The market design dimension can be thought as a sequence of market designs starting from the most demanding in terms of information load and intelligence power of social planner like perfect competition design or Hungarian Algorithm [5,6,19] through less intelligent such as greedy matching defined in 8 in Sect. 3, to more emergent and bottom-up designs not requiring a social planner like in Chamberlin's higgling market [2,3].

The network structure dimension should be thought as network changes of two natures: (1) binary type, i.e. weather possible interactions form the complete graph or not, (2) continuous type, i.e. change in other network characteristics, e.g. size, density, clustering.

The existing literature is primarily focusing on scenarios that do not involve networks, which, to be more precise, is equivalent to the complete graph scenario, for most basic economics model, e.g. perfect competition model [16,22]. For Chamberlin's higgling market, there are established experiment result, see [3], and very preliminary simulation results in [2] presented in no systematic way. Establishing a simulation or even analytical properties of Chamberlin's higgling market is an identified research gap, denoted by the point 1 in Fig. 1, although this point can be, of course, more broadly comprehended. Moreover, embedding Chamberlin's market traders in a social network and assessing network impact is another research gap, denoted by point 3 in Fig. 1. Changing Chamberlin's market design on network into more intelligent market design of greedy matching on network and assessing its impact on the market is another research gap, denoted by point 2 in Fig. 1. All those three research gaps will be addressed, to some limited extent, in the Sect. 4 of simulated results. More thorough analysis including analytical solutions, other market designs, e.g. Double Auction [8,20, 21] and more realistic network structures is the matter of future research.

3   Market Designs on Complete and Sparse Bipartite Graphs

The aim of this section is to delineate three distinct market designs that will be scrutinized in Sect. 4. These include the benchmark perfect competition model, Chamberlin's higgling market (detailed in Sect. 3.1), and the greedy matching approach (elucidated in Sect. 3.2). Furthermore, fundamental concepts pertinent to market design, such as social welfare, market efficiency, and bipartite graphs, will be systematically introduced in Subsects. 3.1 and 3.2.

3.1   Chamberlin's Higgling Market Vs Perfect Competition Model

This subsection defines Chamberlin's higgling market design [3], introduces concepts of social welfare, market efficiency, and the ZIT model for use in Sect. 4. It also evaluates this design using literature on human subject experiments and simulations in comparison to the perfect competition model.

Economics Experiments on Human Subjects. Unlike natural sciences, economics and social sciences are ill-suited for laboratory experiments on economic systems due to prohibitive costs [7]. Economists often resort to abstract and mathematical models for experimentation [12,15]. While macroeconomic experiments are prohibitive, microeconomic experiments have a track record [4,11].

Chamberlin in [3] discusses an experiment where students are divided into sellers (S) and buyers (B). Assume that we have $n \in \mathbf{N}$ agents of each type. Each seller i had a good to sell at a minimum price S_i (this value is sellers *cost*), and each buyer j wanted to buy a good at a maximum price B_j (this value is buyers *willingness to pay*). To simplify further derivations, assume that all values of S_i and B_j are unique. Students negotiated transaction prices p_{ij} such that $S_i \leq p_{ij} \leq B_j$. This ensured non-negative profits (π_i) and consumer surplus (CS_j), as well as non-negative value added from a transaction (i,j) (SW, we also use the term *social welfare* to refer to this value):

$$\pi_i = p_{ij} - S_i \geq 0,$$

$$CS_j = B_j - p_{ij} \geq 0,$$

$$SW(\{(i,j)\}) = \pi_i + CS_j = B_j - S_i \geq 0.$$

Additionally, if multiple transactions take place on the market, then they are denoted as a set T of (i,j) tuples. The social welfare from all transactions in T is defined as follows:

Definition 1. *We define social welfare as:*

$$\boldsymbol{SW}(T) = \sum_{t \in T} SW(\{t\})$$

and scaled social welfare as an average over n agents of each type:

$$\boldsymbol{SSW}(T) = \boldsymbol{SW}(T)/n.$$

Having defined B_j and S_i, we can specify a demand function as $x(p) = |\{j : B_j \geq p\}|$, i.e. the number of buyers willing to buy a good at price p, as well as a supply function $y(p) = |\{i : S_i \leq p\}|$, i.e. the number of sellers willing to sell at price p. Clearly, $x(p)$ is a non-increasing function of p whereas $y(p)$ is non-decreasing. The perfect competition model defines the equilibrium price set $P^* = \{p : x(p) = y(p)\}$. Note that this set is always non-empty as we assumed that all S_i and B_j are distinct. To simplify notation, without affecting further results, we pick one element $p^* \in P^*$ and call it an equilibrium price. The equilibrium volume is defined as $x^* = x(p^*) = y(p^*)$ [22]. Chamberlin constructed demand and supply functions from B_j and S_i respectively and calculated the equilibrium price and volume. Chamberlin's experiments showed that students "traded too much", with actual sales volume and price diverging from equilibrium predictions. This was due to the engagement of extramarginal traders in transactions.

Definition 2 (Intramarginal and extramarginal traders). *The set of intramarginal sellers is defined as $\{i : S_i \leq p^*\}$, while the set of intramarginal buyers is defined as $\{j : B_j \geq p^*\}$. Traders not meeting these conditions are extramarginal.*

While increased trading volume may seem beneficial, the goal for market designers should be to maximize the total social welfare $\boldsymbol{SW}(T)$ from all trades T that took place on the market. Market efficiency is the ratio of the achieved social welfare to the maximum social welfare:

Definition 3 (Market efficiency).

$$\boldsymbol{Eff}(T) = \frac{\boldsymbol{SW}(T)}{\max_{T'} \boldsymbol{SW}(T')},$$

where the maximum in the denominator is taken over all possible trade sets T' that could take place on the market.

Markets with efficiency lower than 100% are called inefficient.

A few relevant properties are presented below to show the relationship between social welfare, market efficiency, prices, and volume. The first observation is that social welfare is independent of prices since social welfare is driven by the pairs of sellers and buyers engaged in transactions, but not particular transaction prices. Note that the transaction price does not enter the formula for social welfare. Next, a well-known economic result is that the equilibrium price denoted as p^* implies maximum social welfare, see [22]. However, price equilibrium is not a necessary condition for the maximum social welfare, as the maximum social welfare is achieved if and only if all intramarginal traders are trading and nobody else. Still, equilibrium volume is a necessary, although not sufficient, condition for maximum social welfare. Since according of trading among intramarginal pairs of sellers and buyers is a necessary condition for maximum social welfare, so is the number of intramarginal pairs of sellers and buyers, which is fixed and equal to the equilibrium volume. As a direct consequence of the above property, one can conclude that trading more than the equilibrium volume destroys social welfare since the equilibrium volume is the necessary condition for the maximum social welfare.

Chamberlain's results [3] showed that excess transactions involving extramarginal trades result in efficiency loss. This inefficiency, due to "too much trading", will be further explored in the next subsection.

Simulation of Zero-Intelligence Traders. The above results of loss efficiency had been already reached by Chamberlin in 1933 in his seminal book of *The Theory of Monopolistic Competition* [2], when he was among the first economists to do the following simulation manually:

Definition 4 (ZIT's Chamberlin bilateral higgling simulation).
Initiate T - the set of trades made so far - to $\emptyset$.
Repeat:

1. *sample the random pair (i^*, j^*) of: (a) a seller $i^* \in \{i : \forall_j (i,j) \notin T\}$ and (b) a buyer $j^* \in \{j : \forall_i (i,j) \notin T\}$ such that neither of them has traded,*
2. *make the pair (i^*, j^*) trade, if $S_{i^*} \leq B_{j^*}$,*
3. *if the pair (i^*, j^*) traded, than update the set: $T \leftarrow T \cup \{(i^*, j^*)\}$*

until no further trade is possible, i.e. $\min_{i:\forall_j (i,j) \notin T} S_i > \max_{j:\forall_i (i,j) \notin T} B_j$.

In this paper, the decision-making behaviour described in Definition 4, in which there is no bargaining and a transaction takes place as soon as it satisfies a necessary condition of $S_i \leq B_j$, is called *Zero-Intelligence Traders (ZIT)*. The name was coined in [8], in which ZIT simulation was employed to evaluate another market design called Double Auction. Experimental evaluation of this market design is reported in [20].

In our study, we will conduct an actual computer simulation for which the above simulation defined in 4 is a special case, since our general model will assume a non-complete bipartite graph of traders. For simulation purposes we will assume that both consumers' willingness to pay and sellers' costs are following identical independent uniform distributions, i.e. S_i, B_j are distributed as iid random variables taken from $U(0,1)$. For this specification, it is easy to show that the maximum social welfare as number of traders $n \to \infty$ is $n/4(1 + o(1))$ with trades taking part for $p^* = 1/2 + o(1)$.

The maximum social welfare of $n/4$ (after ignoring the error term of $o(n)$) will serve as the benchmark for the evaluation of market designs and market structures considered in this and next sections. The market efficiency will measure the social welfare of a particular market design and structure in the relation to the value of maximum welfare.

Below we demonstrate that the market design of Chamberlin's bilateral higgling, as specified in Definition 4, is inefficient, as its market efficiency for identical uniform distributions of willingness to pay and costs as well as number of traders $n \to \infty$ is approximately 73.6%, i.e. approximately 26.4% of value is not achieved.

To show this, we use the following heuristic argument. Let us define the function $X : R^+ \times [0,1] \to [0,1]$ by the following differential equation:

$$\frac{\partial}{\partial t}X(t,v) = -X(t,v)\int_0^{1-v} X(t,z)\,dz,$$

$$X(0,v) = 1 \quad for \quad v \in [0,1].$$

$X(t,v)$ denotes the probability that at time t, a seller (or a buyer) with value v (or $1-v$, respectively) has not yet traded. At the beginning of the process, i.e. $t = 0$, all traders ($v \in [0,1]$) have not yet traded ($X(0,v) = 1$). Instantaneous decrease of this probability, i.e. $\frac{\partial}{\partial t}X(t,v)$, is proportional to the amount of sellers who have not yet traded, i.e. $X(t,v)$, and the amount of buyers who have not yet traded and would be willing to do so, i.e. $\int_0^{1-v} X(t,z)\,dz$.

By monotonicity, it is clear that $\bar{X}(v) = \lim_{t\to\infty} X(t,v)$ exists. $\bar{X}(v)$ gives the limiting probability (as $n \to \infty$) that a seller with value v does not trade, which is also the probability that a buyer with value $1-v$ does not trade. The total fraction of participants who trade is given by:

$$TP = 1 - \int_0^1 \bar{X}(v)\,dv.$$

$\int_0^1 \bar{X}(v)\,dv$ denotes the average non-participation fraction (averaged over all traders with v distributed uniformly), so TP is the average market participation. The total scaled social welfare **SSW** is:

$$\begin{aligned}
\textbf{SSW} &= \int_0^1 v(1-\bar{X}(1-v))dv - \int_0^1 v(1-\bar{X}(v))dv \\
&= \int_0^1 (1-v)(1-\bar{X}(v))dv - \int_0^1 v(1-\bar{X}(v))dv \qquad (1) \\
&= \int_0^1 (1-2v)(1-\bar{X}(v))dv.
\end{aligned}$$

The last expression denotes the integral of the difference between inverse demand and supply functions, i.e. $(1-v) - v = 1 - 2v$, weighted by the participation rate $(1 - \bar{X}(v))$.

It seems that $\bar{X}(v) = 0$ for $v \leq \frac{1}{2}$, and $\bar{X}(v) \geq 2v - 1$ for $v \geq \frac{1}{2}$ (the actual function has some curvature on the interval $[1/2, 1]$). The intuition is that all intramarginal traders, both sellers and buyers, as defined in Definition 2, do engage in trading, i.e. $\bar{X}(v) = 0$ for $v \leq \frac{1}{2}$. However, we see that also some extramarginal traders do trade, i.e. $\bar{X}(v) \geq 2v - 1$ for $v \geq \frac{1}{2}$. From these bounds we get that **SSW** $\geq \int_0^1 \max(2v - 1, 0)^2 \, dv = 1/6$, whereas the maximum SW is $1/4$.

According to a numerical calculation, it turns out that $\bar{T} \approx 0.71$ (71% of participants trade), and **SSW** ≈ 0.184, which is approximately 73.6% of the maximum social welfare, which constitutes the market efficiency. Conducted simulations for $n = 1,000$ in Sect. 4 confirm this asymptotic argument.

As mentioned above, the trade participation, i.e. fraction of traders engaged in a transaction, is approximately 71%, instead of 50% as it would be in the efficient situation. This 71% can be decomposed among:

– 50% are intramarginal traders,
– remaining $\approx 21\%$ are extramarginal traders, i.e. sellers $\{i : S_i > \frac{1}{2}\}$ and buyers $\{j : B_j < \frac{1}{2}\}$ resulting in "too much trading" and inefficiency.

As a further potential research question would it be interesting to generalize above two propositions and specify conditions regarding the class of demand and supply functions, for which those qualitative and quantitative properties hold.

3.2   Greedy Matching of Traders on Network

The aim of this subsection is to introduce another market design for a bilateral exchange, i.e. greedy matching of traders, in order to compare it to Chamberlin's higgling market from the previous subsection, Subsect. 3.1. The performance of greedy matching algorithm is reported in Sect. 4.

Greedy Matching of Traders on a Complete Graph. Given a significant efficiency loss in Chamberlin's higgling market design discussed in Subsect. 3.1, there is a need to come up with another market design, which would result in an efficiency gain and make it closer to perfect competition model's outcomes. Such mechanism would require so called social planner who would know traders' evaluations, i.e. B_j and S_i and would match traders according to a specified algorithm, resulting in a higher social welfare than in Chamberlin's higgling market. One such algorithm is defined below:

Definition 5 (Greedy algorithm for traders' matching).

1. *sort sellers in non-decreasing order of S_i,*
2. *sort buyers in non-increasing order of B_j,*
3. *match greedily until $B_j < S_i$.*

This is an efficient algorithm of a linear complexity with regard to the number of traders resulting in the maximum social welfare, provided that each pair of traders (i, j) can trade with each other.

Greedy Matching of Traders on a Non-complete Bipartite Graph. An interesting extension of the algorithm in Definition 5 is to imagine that traders are embedded onto a random (in particular, non-complete) bipartite graph, see Definition 7. As a result, they cannot trade with everybody, as it would be the case for the complete bipartite graph, see Definition 6, but only with those traders with whom they have a connection to, i.e. there is an edge between them.

Definition 6 (Complete bipartite graph). *Complete bipartite graph $G = (V, E_{complete})$ is the pair of:*

1. *the set of vertices $V = B \cup S$ such that $B \cap S = \emptyset$, i.e. the union of non-overlapping sets of buyers and sellers,*
2. *the set of all possible connections (edges) between sellers and buyers , i.e. $E_{complete} = \{(i,j) : j \in B, i \in S\} = S \times B$.*

Definition 7 (Random bipartite graph with probability p). *Random bipartite graph with probability $p \in [0,1]$ is defined as $G = (V, E_p)$, i.e. the pair of:*

1. *the set of vertices $V = B \cup S$ such that $B \cap S = \emptyset$, i.e. the union of non-overlapping sets of buyers and sellers,*
2. *the set of connections (edges) between sellers and buyers, which is constructed in a way that for each pair of nodes $(i,j) \in E_{complete}$, we independently introduce an edge (i,j) in E_p with probability p.*

Here is the extension of the algorithm defined in 5 to work in non-complete setting.

Definition 8 (Greedy matching of traders on network).

1. *Initiate the set of trades T to $\emptyset$,*
2. *for each edge $(i^*, j^*) \in E_p$ calculate the value $SW(\{(i^*, j^*)\}) = B_{j^*} - S_{j^*}$,*
3. *sort the pairs (i^*, j^*) of sellers and buyers with respect to $SW(\{(i^*, j^*)\})$, in a non-increasing order,*
4. *iterate over sorted pairs (i^*, j^*):*
   (a) if neither of them has traded, i.e. $i^* \in \{i : \forall_j (i,j) \notin T\}$ and $j^* \in \{j : \forall_i (i,j) \notin T\}$ as well as $S_i \le B_j$ then make the pair (i^*, j^*) trade and update the set: $T \leftarrow T \cup \{(i^*, j^*)\}$
   until no further trade is possible, i.e. $S_{i^*} > B_{j^*}$.

Since the most time consuming part of the aforementioned algorithm is sorting the $n^2 p$ pairs of sellers and buyers, the computational complexity of the algorithm is $O(n^2 p \log(n^2 p)) = O(n^2 p \log(n))$, where n denotes the total count of both sellers and buyers. Clearly, one may implement a faster algorithm but we do not need it for our experiments.

Greedy algorithm defined in 8 does not guarantee to find a global optimum and maximise the social welfare for a given bipartite graph, unless a graph is complete, as stated before. However, the global optimum can be achieved by

Hungarian algorithm [6] which is much more costly numerically and prohibitive for large bipartite graphs, so it will not be evaluated in this paper. The performance of the algorithm from Definition 8 is reported in Sect. 4.

Below, we provide a comprehensive summary of various market designs. We focus on four designs in particular, two of which are compared in the results section, and two others that are worth mentioning:

- **Chamberlin's Bilateral Haggling ZIT market** (as defined in Definition 4) serves as our benchmark model. This model is simplistic, assuming no top-down intervention and only bottom-up emergent behaviours of traders. It does not attribute any intelligence to traders in terms of learning, information collection, bargaining capabilities, etc. Consequently, as demonstrated in Sect. 4 and in the literature [3], this market design suffers from losses in social welfare and market efficiency,
- The **Hungarian Algorithm** is designed to identify market-clearing prices that maximize social welfare within a given bipartite graph, see [5,6,19]. This is achieved by enabling a regulator to dictate the trading partners. Despite its advantages, the algorithm is characterized by high computational complexity of $O(n^3)$.
- **Greedy Matching** (as defined in Definition 8) is our proposed efficient heuristic algorithm. It stands between the two extremes in terms of both computational requirements and resulting market efficiency. Similar to the Hungarian Algorithm, it assumes a market regulator making top-down decisions about who trades with whom. As presented in Sect. 4, greedy matching exhibits high market efficiency.
- **Double Auction** should be positioned between ZIT's Chamberlin bilateral haggling and our greedy matching algorithm. It allows for some local information exchange, but not to the global extent as in our greedy algorithm. It assumes bottom-up trader behaviour with the institution of public price quoting, facilitating information exchange. High efficiency of this market design for complete bipartite graphs has been demonstrated in [8,20], but there is still no evidence in the literature for sparse bipartite graphs.

4   Simulation Results

This section is dedicated to addressing the research gaps identified and outlined in the research agenda depicted in Fig. 1. Specifically, it will showcase exploratory simulation results, with a focus on two fundamental characteristics of markets:

1. market efficiency (Subsect. 4.1),
2. trade participation (Subsect. 4.2).

Those two metrics will be differentiated based on three main drivers: average degree of trading network, market design and network size. The presented outcomes of market efficiency and trade participation are calculated as the average out of 1000 simulations.

For each simulation, traders are embedded in a newly-generated graph. Since traders are of two types, either seller or buyer, and there is no value from connections within sellers or within buyers as they can make transaction only between (not within) them, it is natural to employ bipartite graph to model interactions between agents, see [6,14]. In this study, random bipartite graph is generated according to the Definition 7.

The Julia code employed for this simulation experiment is available at GitHub[1].

4.1   Market Efficiency Drivers

Figure 2 demonstrates how the market efficiency depends on:

- average degree of bipartite graph, i.e. average number of acquainted traders, with whom an average trader is connected to, $np \in [0.01, 1000]$,
- market design algorithm of decision making, either Zero-Intelligence Traders or greedy matching of traders,
- market size, i.e. the number of sellers and buyers, $n \in \{10, 100, 1000\}$.

In the following subsections we describe the impact of each of these drivers separately based on Fig. 2.

Average Degree of Trading Network. Average degree is the expected average number of acquainted traders $(= np)$, but it is also a partial measure of network density, i.e. p, the probability of edge existence. Based on Fig. 2a, the average degree comes as the major driver of market efficiency. The basic observation is that the larger the average degree, the higher market efficiency for both market design decision mechanisms of ZIT and the greedy matching of traders regardless of the number of traders.

The impact of average degree on market efficiency is not constant, but most significant for degrees between 0.1 and 10, accounting for approximately 90% of the change. This is evident from the steepest ascent of curves in Fig. 2b. Degrees less than 0.1 or greater than 10 have minimal effect on market efficiency. An average degree of 5 and 10 achieves roughly 89% and 95% of potential market efficiency[2], respectively. This suggests that moderately dense markets with only 5 or 10 acquainted traders can achieve high potential market efficiency, which has significant implications for policy makers and market designers. From a traders' perspective, they should aim to have at least 5 acquainted traders to achieve roughly 89% of potential profits or consumer surplus.

---

[1] https://github.com/Matzawisza/TradeInNetwork.

[2] Potential market efficiency is the maximum market efficiency in a complete graph for a given market design, i.e., it is roughly 73.6% for ZIT and 100% for a greedy matching of traders.

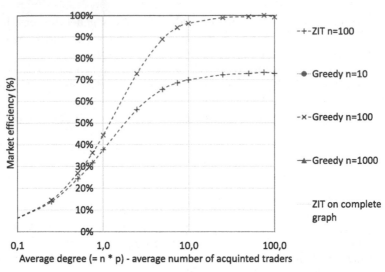

(a) Superiority of greedy matching over ZIT ($n = 100$)

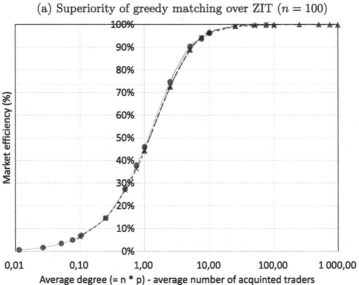

(b) No market size impact for greedy matching

Fig. 2. Market efficiency comparison between ZIT and greedy matching for varying average degree ($np \in [0.01, 1000]$) and market size ($n \in 10, 100, 1000$) **Source:** Own work

Market Design. Two market designs are considered in Fig. 2a:

– Zero-Intelligence Traders (ZIT) like in the Chamberlin's higgling market,
– greedy matching of traders, as described in Definition 8.

The choice of market design has the enormous impact on the actual and potential market efficiency. The choice is especially crucial for markets with the average degree of at least 1, so nearly for all markets in practice. The market efficiency of a greedy matching design in comparison to ZIT exhibits following properties:

- greedy matching achieves higher market efficiency than ZIT for all (considered in simulations) values of the average degree and market sizes,
- greedy matching is able to achieve nearly 100% market efficiency, if the trading network is dense enough, while the maximum market efficiency of ZIT is only roughly 73.6% for a complete graph.
- the difference between greedy matching and ZIT starts widening significantly from the average degree of 1 and continues this significant widening till 10 of acquainted traders, when it achieves nearly the maximum difference between two designs of roughly 26.4%.

Based on the above properties we come up with following recommendations for policymakers, which depend on the average number of acquainted traders:

- for the average degree of at least 1, the greedy matching is strictly preferred over ZIT market design, as the difference between the two is of significant magnitude, i.e. at least 5 percentage points. The highest consequence of this market design choice happens for the average degree of at least 10, where the difference achieves nearly its maximum of 26.4%,
- for the average degree lower than 1, even though the greedy matching is still better than ZIT, the difference between both market designs is of less significant magnitude (although relatively it might still matter), so the choice of any particular design is of less importance. In this case, a better policy recommendation would be to increase the average degree of a network.

Market Size. Market size denoted by parameter n seems to be a major characteristics of all real-world markets. The intuition goes that the larger market the more trading opportunities and stronger market forces push it to equilibrium state. Therefore, market size is one of tree parameters considered in our simulation and its impact is depicted in Fig. 2b.

The small market of 10 sellers and 10 buyers is marginally better performing in terms of market efficiency than larger markets composed of 100 or 1000 sellers and buyers although, the impact is of negligible size, i.e. maximum recorded difference is 3.6% between $n = 10$ and $n = 100$ for average degree of 10. The further increase of market size from 100 and 1000 does not decrease to the same degree the market efficiency (the maximum difference is 0.6% for the average degree of 75).

The difference of market efficiency due to market size is more pronounced although still negligible for:

- ZIT (the maximum within difference of 3.6%) rather than greedy matching (the maximum within difference of 1.9%) market design,

– the average degree between 1 and 10 (the maximum difference within this range is 3.6% for the average degree of 10) rather than outside this range (the maximum difference outside this range is 1.6% for the average degree of 0.8).

4.2 Trade Participation Drivers

Trade participation, i.e. the total volume of transactions out of all possible transactions, is another metrics of interest to both researchers and policy makers. As discussed in Subsect. 3.1, the equilibrium volume is a necessary condition for the market efficiency of 100%. For the parametrization of our simulation, the equilibrium volume is equivalent to the trade participation of 50%. Hence, trading above or below this level will imply no market efficiency. The dependence of trade participation on: (1) the average degree of trading network, (2) market design algorithm of decision making, and (3) market size is depicted in Fig. 3 and discussed in following subsections.

Average Degree of Trading Network. Impact of the average degree of trading network on trade participation is positive, as indicated in Fig. 3a, although the impact is not constant, but resembles logistic shape relationship.

For low values of the average degrees, up to 1, the participation is increasing from roughly 0.5% for the average degree of 0.01 till roughly 30% of trade participation for the average degree of 1 for the market design of either ZIT or greedy matching. The low trade participation is due to the sparse trading network with many traders having no connections to other traders and not being able to make any transactions. Also, those traders who have 1 or a few more connections (but not many) might find themselves unable to make a trade, if neither of possible trading pair can generate a social welfare-improving transaction. Hence, the probability of trading is low.

For medium values of average degrees between 1 and 10, the trade participation is increasing further and achieving completely (for greedy matching) or nearly (for ZIT) its maximum level at average degree of 10. However, this maximum level of participation is different for each market designs, which will be discussed in the next subsection.

For large values of the average degrees, higher than 10, the ZIT market design still grow to achieve its maximum level of roughly 71%, while greedy matching is stable at the level of 50%.

The average degree can be as low as 2.5% to achieve 87.2% of maximum participation market rate for greedy matching and only 5 acquainted traders for achieving 97.5% of maximum trade participation.

Market Design Algorithm of Decision Making. Market design choice is a major driver of the trade participation, especially for the average degree higher than 10, see Fig. 3a. For the average degree lower than 1 both considered market designs do not differ in terms of trading participation. The difference starts

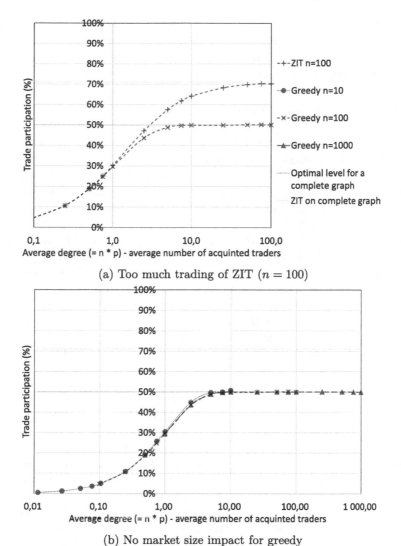

(a) Too much trading of ZIT ($n = 100$)

(b) No market size impact for greedy

Fig. 3. Trade participation comparison between ZIT and greedy matching for varying average degree ($np \in [0.01, 1000]$) and market size ($n \in 10, 100, 1000$) **Source:** Own work

widening from the average degree of roughly 1 on and achieves nearly its maximum discrepancy at and above 10 of acquainted traders, which is 21%.

The greedy matching algorithm achieves optimal participation rate of 50% at the average degree of 7.5%. The ZIT design, as we know already, results in "too much trading" and the trade participation of roughly 71% for more dense networks of 100 trading partners for each trader.

Market Size. Analogously as for market efficiency, also here the market size is of negligible impact, see Fig. 3b. The largest impact of market size is noticed when comparing ZIT designs of 10 and 100 traders. It turns out that smaller market exhibits higher trade participation of 2.2%.

5   Conclusions and Further Research

This paper investigated the effects of market design, market size, and trading network structure on market efficiency and trade participation rate. Two market designs were considered: Zero-Intelligence Traders (ZIT) in Chamberlin's bilateral higgling market and a greedy matching of traders. Both sellers and buyers were embedded in a random bipartite graph with varying network density. Market sizes ranged from 20 to 2000 traders.

Simulations showed that greedy matching outperforms ZIT for non-sparse trading networks. Market efficiency can be significantly improved for both market designs by increasing the average degree of trading networking from 1 to 5 or 10, enabling greedy matching to achieve 89% and 95% of its maximum efficiency, respectively. Contrary to popular belief, market size had no significant impact.

Our model, presented in this paper, offers a simplified representation of traders in a bipartite random graph. However, it is important to acknowledge that real-world scenarios are often more intricate and dynamic. Our model does not incorporate several features observed in real-world markets, such as the qualities of ties among traders, e.g. strong or weak ties, degree distribution of bipartite graphs exhibiting power-law distributions, phenomena such as "rich get richer", communities, assortativity, homophily, aversion, and the dynamic nature of such markets. These elements introduce complexities that our model may not fully capture. Despite these limitations, the authors conjecture—though not explicitly proven or demonstrated—is that the inclusion of these more realistic network features would primarily impact the quantitative results of our model, while leaving the qualitative outcomes relatively stable. This conjecture is based on fact that the qualitative results are usually more robust than quantitative ones. However, it is crucial to interpret these results within the context of these inherent complexities and limitations. Further research is needed to validate this assumption and to quantify the potential impact of these real-world features on our model's outcomes.

Our conducted simulations served as a tool to pinpoint intriguing issues. We are now shifting our focus towards proving theorems about the processes that were simulated. Looking ahead, our research will aim to evaluate the possibility of obtaining analytical results for some manageable models of network, market design, and decision process, offering a contrast to simulation results. This future work will provide a more rigorous understanding of the systems under study. Besides that, the conducted study did not fully exploit the research agenda outlined in Fig. 1. Future investigations could include:

- Enriching the network structure beyond a random bipartite graph to exhibit more realistic features of real-world networks, such as "rich get richer" [1],

"birds of a feather flock together", "six degrees of separation", and small-world property [23] or with community structure [13,18],
- Evaluating the impact of network structure on Double Auction design.

References

1. Barabási, A.L., Albert, R.: Emergence of scaling in random networks. Science **286**(5439), 509–512 (1999)
2. Chamberlin, E.: The Theory of Monopolistic Competition. Harvard Economic Studies ... vol. XXXVIII, Harvard University Press, Cambridge (1933). https://books.google.pl/books?id=imDGxAEACAAJ
3. Chamberlin, E.H.: An experimental imperfect market. J. Polit. Econ. **56**(2), 95–108 (1948)
4. Davis, D.D., Holt, C.A.: Experimental Economics. Princeton university Press, Princeton (2021)
5. Demange, G., Gale, D., Sotomayor, M.: Multi-item auctions. J. Political Econ. **94**(4), 863–872 (1986)
6. Easley, D., Kleinberg, J., et al.: Networks, Crowds, and Markets. Cambridge Books, Cambridge (2012)
7. Friedman, D., Sunder, S.: Experimental Methods: A Primer for Economists. Cambridge University Press, Cambridge (1994)
8. Gode, D.K., Sunder, S.: Allocative efficiency of markets with zero-intelligence traders: market as a partial substitute for individual rationality. J. Polit. Econ. **101**(1), 119–137 (1993)
9. Jackson, M.O., Rogers, B.W.: Meeting strangers and friends of friends: how random are social networks? Am. Econ. Rev. **97**(3), 890–915 (2007)
10. Jackson, M.O., et al.: Social and Economic Networks, vol. 3. Princeton University Press, Princeton (2008)
11. Kagel, J.H., Roth, A.E.: The handbook of Experimental Economics, vol. 2. Princeton University Press, Princeton (2020)
12. Kamiński, B.: Podejście wieloagentowe do modelowania rynków: metody i zastosowania. Szkoła Główna Handlowa, Oficyna Wydawnicza (2012)
13. Kamiński, B., Prałat, P., Théberge, F.: Artificial benchmark for community detection (ABCD)-fast random graph model with community structure. Netw. Sci. **9**(2), 153–178 (2021)
14. Kaminski, B., Prałat, P., Théberge, F.: Mining Complex Networks. CRC Press, Boca Raton (2021)
15. Law, A.M.: Simulation Modeling & Analysis, 5th edn. McGraw-Hill, New York (2015)
16. Mas-Colell, A., Whinston, M.D., Green, J.R., et al.: Microeconomic Theory, vol. 1. Oxford University Press, New York (1995)
17. Newman, M.: Networks. Oxford University Press, Oxford (2018)
18. Newman, M.E., Girvan, M.: Finding and evaluating community structure in networks. Phys. Rev. E **69**(2), 026113 (2004)
19. Pass, R.: A Course in Networks and Markets: Game-theoretic Models and Reasoning. MIT Press, Cambridge (2019)
20. Smith, V.L.: An experimental study of competitive market behavior. J. Polit. Econ. **70**(2), 111–137 (1962)

21. Smith, V.L.: Effect of market organization on competitive equilibrium. Q. J. Econ. **78**(2), 181–201 (1964)
22. Varian, H.R.: Microeconomic analysis, vol. 3. Norton New York (1992)
23. Watts, D.J., Strogatz, S.H.: Collective dynamics of 'small-world'networks. Nature **393**(6684), 440–442 (1998)

Network Embedding Exploration Tool (NEExT)

Ashkan Dehghan[1]($\boxtimes$), Paweł Prałat[1], and François Théberge[2]

[1] Department of Mathematics, Toronto Metropolitan University, Toronto, ON, Canada
{ashkan.dehghan,pralat}@torontomu.ca
[2] Tutte Institute for Mathematics and Computing, Ottawa, ON, Canada
theberge@ieee.org

Abstract. In this paper, we introduce **NEExT**(**N**etwork **E**mbedding **Ex**ploration **T**ool) for embedding collections of graphs via user-defined node features. The advantages of the framework are twofold: (i) the ability to easily define your own interpretable node-based features in view of the task at hand, and (ii) fast embedding of graphs provided by the **Vectorizers** library. In this exploratory work, we demonstrate the usefulness of **NEExT** on collections of synthetic and real-world graphs.

1 Introduction

Many real-world as well as artificial systems and processes can be represented as graphs. Such systems include social networks, financial transactions, supply chains and molecular structures, to name a few. In many cases, one needs to consider a collection of related graphs, whether they are the result of multiple systems (e.g., different proteins) or are produced by a dynamic process of the same network (e.g., evolution of a social network, graphs induced by different communities, or ego-nets around various nodes). A significant challenge in most scenarios is the absence of ground-truth labels for graphs and nodes, and therefore one needs to use unsupervised techniques to study various properties of such networks. To address these challenges, we introduce an **N**etwork **E**mbedding **E**xploration **T**ool (**NEExT**)[1], which can be used to analyze collection of graphs in an unsupervised fashion.

Node embedding is a transformation of nodes of a network into a set of vectors [1]. Due to their spectacular successes in various applications, they are becoming increasingly popular in the ML community. There are over 100 algorithms available to use and frameworks to evaluate them (such as [8]). Independently, many analytic tasks (such as classification, clustering, and regression) in various domains, including social networks, cybersecurity, bio- and chemo-informatics, require representing graphs as fixed-length feature vectors [1]. For example, embeddings of graphs representing program's calls could be used to detect malware [18], embeddings of graphs representing chemical compounds

---

[1] https://pypi.org/project/NEExT/.

M. Dewar et al. (Eds.): WAW 2024, LNCS 14671, pp. 65–79, 2024.
https://doi.org/10.1007/978-3-031-59205-8_5

could be used to predict properties of the associated compounds such as solubility and anti-cancer activity [19,25].

Historically, graph kernels were considered to be a standard way to deal with the above graph analytics tasks. In this approach, the similarity (kernel value) between pairs of graphs is computed by recursively decomposing them into simpler substructures (such as random walks, shortest paths, graphlets) and defining similarity (kernel) between these substructures. After that, some standard kernel methods, such as Support Vector Machines (SVMs), can be used to classify or cluster graphs. Note that many algorithms of this nature do not explicitly produce graph embeddings and so they cannot be immediately used for general ML tasks.

To overcome this limitation, another powerful technique was introduced recently in the literature [16,22]. We start with extracting features of nodes of a graph G via some node embedding algorithm. Such cloud of n points, corresponding to vectors of features of n nodes of G, can be easily normalized so that it can be viewed as the probability distribution on a metric space equipped with a distance, such as the Euclidean distance. Then, the Wasserstein distance can be used to measure the distance between two graphs by computing the distance between the two corresponding probability distributions. The Wasserstein distance is a metric and is linked to the optimal transport problem [23] which aims to find an optimal way to transport the probability mass associated with one graph to the one associated with another one. This distance is sometimes referred to as the earth mover's distance since in 2-dimensional case one can think of it as moving piles of dirt. Finally, some algorithm is used to embed graphs into k-dimensional space of vectors such that the Wasserstein distance between graphs matches the distance between the corresponding vectors as much as possible. In this paper, we introduce a framework that builds on ideas from [16,22]. On top of providing an efficient and user-friendly exploratory networks analysis tool, our main contribution can be summarized as follows.

- The framework not only utilizes a number of standard classical and structural node embeddings but allows to include hand-crafted, user-defined feature vectors that, for example, measure the distribution of power (by including Pagerank or some other centrality measures) or expansion of ego-nets around nodes. This approach has a few immediate benefits: it is much faster to compute such features than embed all nodes, the results are interpretable and more robust.
- The framework utilizes various techniques and metrics for approximating the distances between graphs such as Wasserstein distance and Sinkhorn algorithms, in addition to a computationally efficient approximate Wasserstein vectorization approach. These tools are available and maintained in the **Vectorizers** package.
- For supervised learning (when labels for graphs are available), the framework selects (in an automated way) a subset of available node features and appropriately normalizes them for the best outcome of a given ML supervised task

at hand such as classification or regression. (Work in progress, not discussed in this proceeding version of the paper.)

2   The Framework

Consider a collection of graphs $G_1, G_2, \ldots, G_m$. Our approach follows the following steps (details are provided in the following subsections):

1. Pre-process the collection of graphs.
2. For each graph G_i with n_i nodes, build k-dimensional vector representations for all the nodes.
3. Given m collections of k-dimensional vectors, one collection for each graph, compute d-dimensional embedding of the graphs via either the `Wasserstein`, `Sinkhorn`, or `ApproximateWasserstein` distance.

Note: In the following description, individual components of the graph collection ($G_i s$), will be referred to as *subgraphs*.

2.1   Pre-processing

In the pre-processing layer, **NEExT** loads each subgraph into a **GraphCollection** object, assigning to each subgraph appropriate details such as graph labels, graph statistics, etc. In this layer, we can also filter each subgraph for its largest connected component.

2.2   Vectorizing the Nodes

Several methods to obtain vector representations for the nodes of each graph are available in the framework. However, one advantage of **NEExT** is that it is easy to add other vector representations for the nodes, which can be interpretable and designed specifically for the types of graphs one wants to analyze and downstream Machine Learning task using the generated graph embedding. It can also be used to easily test various families of node-based features in order to select suitable ones. In the final implementation of **NEExT** (work in progress), if labels are available for some graphs, then such selection can be done in a supervised way. Node features can be computed recursively up to some maximum value k (resulting a k dimensional feature vector) by averaging its values for neighbours up to j hops away for $1 \leq j \leq k$. Moreover, features can be concatenated to obtained even higher dimensional representations.

LSME. One of the built-in structural embedding algorithms is called **Local Signature Matrix Embedding (LSME)**. This technique uses a random-walk algorithm to capture local structural properties of nodes. The algorithm results in a k-dimensional vector, where each element measures the transition probability between various neighbourhoods around a given node[2].

---

[2] https://github.com/ashdehghan/LSME.

Centrality Measures. We compute various commonly used centrality measures such as **PageRank**, **Closeness Centrality**, **Degree Centrality**, and **Eigenvector Centrality**. Details for those measures can be found in, for example, [11].

Expansion Properties. For each node v, let $\hat{m}_i$ be the number of neighbours at distance i from v, for $i \in \{1, 2, \ldots, k\}$. Our goal is to embed vertices of possibly different degrees that expand in a similar way close to each other. Hence, we consider the following feature vector for a node v:

$$E(v) = \left(\frac{\hat{m}_1}{1 \cdot \bar{d}}, \frac{\hat{m}_2}{n_1 \cdot \bar{d}}, \cdots, \frac{\hat{m}_k}{n_{k-1} \cdot \bar{d}} \right),$$

where $\bar{d} = \frac{1}{N} \sum_{v \in V} \deg(v) = \frac{2|E|}{N}$ is the average degree. For good expanders, one would get a collection of vectors that are close to $(1, 1, \ldots, 1)$.

2.3 Embedding of the Graphs

Assume that we have a collection of graphs $G_i, 1 \leq i \leq m$, with respectively n_i nodes, and a k-dimensional vector representation for each node. Then, each graph can be seen as a distribution of points over k-dimensional space, and we can use some measure of distance between distributions to embed the graphs in some vector space. One possible approach is to use the Wasserstein distance, which is obtained by finding the optimal transport plan between distributions; this is also known as the earth mover's distance between distributions (i.e. measure the amount of "work" to move mass from one distribution to the other); see, for example, [22]. Embedding graphs this way is similar to the context of document embedding[3], where each word is represented by a vector (obtained via some word embedding algorithm), and each document is a "bag of word vectors".

Given m graphs, computing all such distances requires estimating $O(m^2)$ pairwise distances, which has a high computational cost. One solution to this issue is to define some reference distribution (for example via averaging the vectors), and find the optimal transport plan from each graph's "bag of vectors" to this reference distribution. This is know as linear optimal transport (LOT) [23], which is used, for example, in [16]. We use the implementation of this approach from the easy to use and frequently maintained **Vectorizers**[4] Python package, which solves the LOT and computes embeddings by computing the SVD (Singular Value Decomposition) of the optimal transport plans.

Computing the Wasserstein distances, even using a reference distribution, can still be prohibitive for some large problems. We therefore consider two faster methods which are also implemented in **Vectorizers**. The first one uses the Sinkhorn distance which is based on entropic regularization of the transport plans; see [5]. The other one, ApproximateWasserstein, also solves the LOT but using a single-point reference distribution obtained via averaging, as described

---

[3] https://vectorizers.readthedocs.io.
[4] https://pypi.org/project/vectorizers/.

in [2]. Embeddings are obtained using SVD with scaling according to the singular values.

As a rule of thumb, when $k \ll m$, one can use the Wasserstein or Sinkhorn approach while for larger k, the ApproximateWasserstein can be used for better performance.

3   Experiments

We illustrate the use of our framework to analyse synthetic as well as real-life networks. The goal is to explore various capabilities of the framework for both unsupervised and supervised applications. In the first subsection, we use the *Artificial Benchmark for Community Detection* (**ABCD**) framework [10], to generate synthetic graphs. The **ABCD** framework is a random graph model framework with community structure and power-law distribution for both degrees and community sizes. The goal here is to explore and highlight various properties of our framework in a controlled environment and showcase its use from a practitioners point of view. Therefore, we consider idealized and synthetically generated cases, while considering more real-world scenarios in the following subsection.

3.1   Synthetic Graphs

The **ABCD** model, an alternative approach to the **LFR** model [17], allows us to generate random graphs with control over power-law distribution of both node degrees and of community sizes, fraction of outlier nodes, noise, and other parameters. We leverage the Julia implementation[5] to generate the synthetic graphs used in this section. However, let us mention that for generating enormous graphs there exists a faster implementation[6] available that uses multiple threads (**ABCDe**) [15].

Undirected variant of **LFR** and **ABCD** produce graphs with comparable properties but **ABCD/ABCDe** is faster than **LFR** and can be easily tuned to allow the user to make a smooth transition between the two extremes: pure (disjoint) communities and random graph with no community structure. Moreover, it is easier to analyze theoretically—for example, in [3,9] various theoretical asymptotic properties of the **ABCD** model are investigated including the modularity function and self-similarities of the ground-truth communities. More importantly, the model is extremely flexible and allows to include outliers [12] (**ABCD+o**) or generate hypergraphs [13] (h–**ABCD**).

To explore various properties of our framework, we designed three experiments. The parameters used in each experiment are outlined in Table 1.

Experiment 1 – Varying Level of Noise. In the first experiment, we explore the effect of noise that is controlled by parameter ξ in the **ABCD** model. We designed this experiment to mimic a dynamic property of a graph in which

---

[5] https://github.com/bkamins/ABCDGraphGenerator.jl.

[6] https://github.com/tolcz/ABCDeGraphGenerator.jl/.

Table 1. ABCD Synthetic Graph Parameters. (The number of nodes is equal to n. The degree distribution follows power-law with exponent γ, minimum δ and maximum Δ. The distribution of community sizes follows power-law with exponent β, minimum c and maximum C. The level of noise is controlled by ξ. Finally, the number of outliers is equal to o. For a more in depth description of each parameter and how it is utilized, we refer the reader to the GitHub repository.)

Parameter	Experiment 1	Experiment 2	Experiment 3
n	200	$\{200, 250, \ldots, 400\}$	200
γ	3	3	3
δ	5	5	5
Δ	10	10	10
β	2	2	2
c	10	10	10
C	20	20	20
ξ	$\{0.1, 0.101, \ldots, 0.9\}$	$\{0.1, 0.2, \ldots, 0.5\}$	0.2
o	0	0	$(150 \times 10) + (150 \times 50)$

an incremental change in a property of a parameter in the graph results in the change in underlying graph structure. In a real world network, this could resemble the change in the polarization of a network in which the boundaries between communities slowly vanish. Of course, we have to note that our synthetic experiment is an idealized version of such system.

We construct 801 graphs, with ξ ranging from 0.1 to 0.9 in steps of 0.001. In the **ABCD** model, ξ controls the fraction of edges that fall into the background graph (almost all of these edges are between nodes from different communities). A sample of four graphs from the 801 generated ones are shown in Fig. 1. We then use ξ values as a label for each graph to be used for a regression task. Steps of this experiment are as follows:

- Generate a collection of graphs with varying level of noise controlled by ξ.
- Generate feature vectors of size k=2 to k=8 for each graph.
- Use the features vectors to construct graph embeddings of dimension $d = 2$.
- Use the graph embedding vectors as features for a regression task to predict the ξ values for each graph.

Once the graph collection is generated and loaded into our framework, we can compute various graph properties on each graph in the collection. For this experiment, we compute the following graph features: **Expansion, LSME, PageRank, Closeness Centrality, Degree Centrality**, and **Eigenvector Centrality**. For each feature, we construct a k dimensional vector. For example, for **PageRank** as a feature with $k = 3$, for each node v we calculate the **PageRank** value of v as well as the average **PageRank** values of neighbours at distance i from v, where $i \in \{1, 2, \ldots, k-1\}$. We also construct larger feature vectors by concatenating the vectors from multiple features. For example, a feature vector of **Expansion** + **LSME** with $k = 3$ is a concatenation of a 3 dimensional **LSME** feature vector and 3 dimensional **Expansion** feature vector, resulting in a 6 dimensional global feature vector.

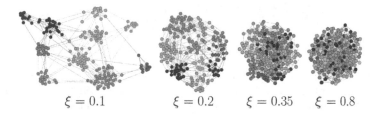

$$\xi = 0.1 \qquad \xi = 0.2 \qquad \xi = 0.35 \qquad \xi = 0.8$$

Fig. 1. Examples of graphs generated using the **ABCD** synthetic graph for Experiment 1, as detailed in Table 1, for $\xi \in \{0.1, 0.2, 0.35, 0.8\}$.

Having defined the feature generation process, we construct features of lengths k from 2 to 8 based on the above list. We then use the approximate Wasserstein technique to embed each graph in the collection into a two dimensional embedding. We chose embedding dimension $d = 2$, since the approximate technique has an upper limit of k for the dimension of the embedded space and the smallest feature vector size is of dimension $k = 2$. Moreover, since our graph embeddings are used in a downstream supervised regression task, we wanted to keep the dimensionality of the embeddings the same to standardize the comparison of the models.

In Fig. 2 we show the two dimensional embedding of graphs build using the **Expansion, LSME**, and **PageRank** features. In all three cases, the underlying feature vectors have length $k = 5$. Each data point (graph embedding) is then coloured based on the available label (ξ). It is clear that in all three cases there is a relation between the value of ξ and the graph embedding vector. To explore this relationship further, we use a regression model for graph embeddings that are built using various graph features.

Using the two dimensional graph embeddings of the graphs, we train regression models using XGBoost[7] to predict the value of ξ for the unlabeled graphs. In our experiment, the train/test split is set to 70/30 and we repeat each experiment 100 times to arrive at the average mean-absolute-error and the standard deviation over the runs, shown as error bars in Fig. 3. Here, the x-axis corresponds to k, the length of feature vectors computed for nodes of each graph, and different colours correspond to combination of various types of features.

We start by highlighting the fact that the overall performance of the models increases (the mean-absolute-error decreases) as the length of the underlying feature vectors increases. This is expected, since increase in k corresponds to a larger window for capturing structural properties of the underlying graph. We

---

[7] https://xgboost.readthedocs.io/en/stable/.

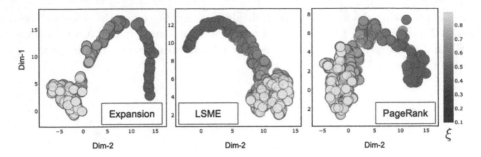

Fig. 2. Two dimensional graph embeddings built using the approximate Wasserstein technique and graph features built using the **Expansion, LSME,** and **PageRank** metrics. The dimension for all the above node embeddings is set to $k = 5$.

show that this trend continues until the length of the feature vectors reaches the diameter of the graphs. At this length scale, the feature vectors are capturing global structural properties of the graph. We note that the diameter for our synthetic graphs is relatively small, since we consider graphs of size $n = 200$ that are relatively good expanders. Lastly, we note that models built on combination of features perform the best, since each feature captures a different structural property of the graph. We note that in this experiment we are not interested in fine-tuning the XGBoost models to achieve the best performance, but rather to illustrate the predictive power of various graph features and associated graph embeddings.

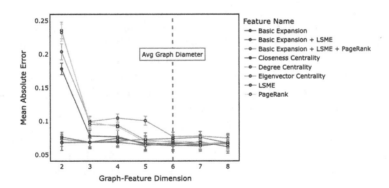

Fig. 3. Mean-absolute-error measured for a regression model built to predict ξ in Experiment 1, as defined in Table 1. The x-axis is the length of the feature vectors computed on each graph. The final graph embedding is uniformly set to $d = 2$.

Experiment 2 – Varying Network Size. In the second experiment, we explore the ability of our framework to capture structural similarities in collections of graphs. In real systems, it is often important to identify structurally similar graphs, regardless of the size of the network. This is often seen in self-similar systems, such as social networks, where particular property presents itself at different scales [3, 20]. To study this effect, we use the **ABCD** model to generate structurally similar networks of various sizes. We achieve this goal by tuning two parameters: the level of noise (ξ) and the number of nodes in each graph (n). As highlighted in Table 1, we build a collection of 25 graphs with $\xi \in \{0.1, 0.2, 0.3, 0.4, 0.5\}$ and $n \in \{200, 250, 300, 350, 400\}$. As it was done in Experiment 1, we compute node features for each graph and use the approximate Wasserstein technique to build an embedding vector for each graph. We fix the feature vector size to $k = 4$ and graph embedding size to $d = 4$. To visualize the final embeddings, we use UMAP[8] to map the final graph embedding into two dimensions.

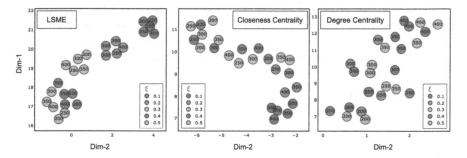

Fig. 4. Two dimensional representations of the approximate Wasserstein graph embeddings built using **LSME, Closeness Centrality**, and **Degree Centrality** graph features. Colours correspond to different values of noise (ξ) and the size of the underlying graphs (n) are shown inside each data point.

In Fig. 4, we show the two dimensional representations of the graph embeddings, coloured and annotated using ξ and n, respectively. In the first chart (left), the underlying feature vector was computed using the **LSME** structural embedding algorithm. This technique captures structural properties of nodes within each graph. Using **LSME** as features, the graph embeddings captures similarities between structural properties of each graph. This can be seen in the final two dimension representation of the embeddings, since graphs with similar structure (ξ value, colour-coded) are clustered together. It is interesting to observe that the quality of the clusters (tightness) decreases as the noise (ξ) increases.

Similar behaviour is also captured by a more simple structural feature, **Closeness Centrality**. We can see in the middle chart in Fig. 4 that **Closeness Centrality** also groups graphs of similar property together, but with lower quality compared to **LSME**. The decrease in clustering quality as a function of the level of noise is more evident in this case. The important observation in these cases points at the fact that the approximate Wasserstein technique using

---

[8] https://umap-learn.readthedocs.io/en/latest/.

LSME or **Closeness Centrality** preserves structural properties of the embedded graphs such as level of noise.

Experiment 3 – Outlier Detection. In this experiment, we control the fraction of outlier nodes in the graph. We construct two sets of graphs. In the first one, 5% of nodes are outliers and in the second group there are more outliers, namely, 25%. For each group, we generate 150 graphs. It is important to note that, since the **ABCD** model is a *randomized* graph generator, each graph in their respective groups are different due to a random nature of the model. This is achieved by setting different random seeds while generating the graphs.

We compute various graph features on each subgraph and use them to construct graph embeddings using the approximate Wasserstein techniques. Here, we consider 9 different sub-groups of features, as defined in Table 2. Each feature type has a dimension of $k = 4$ and we combine different features by concatenating their feature vectors. We note that in this experiment, we train binary classifiers using XGBoost classifier with 70/30 train/test split, and repeat each experiment 100 times to collect enough statistics for model performance.

Table 2. ABCD outlier classification model.

Model	Details
M-0	Expansion
M-1	LSME
M-2	PageRank
M-3	Degree Centrality
M-4	Closeness Centrality
M-5	Eigenvector Centrality
M-6	Expansion, LSME
M-7	Expansion, LSME, PageRank
M-8	Expansion, LSME, PageRank, Degree Centrality, Closeness Centrality, Eigenvector Centrality

In Fig. 5 we show, on the left, the performance of binary-classifiers built for each model (M-0 to M-8) measured using accuracy and (right) a two dimensional clustering of the graph embeddings built using M-8 features. Starting with the right figure, it is clear that the two dimensional representations of graph embedding vectors form two well separated clusters. In an unsupervised setting, one could use a technique such as DBSCAN [7] to identify these two clusters, even if the underlying classes are not known. In this experiment however, we know that the underlying graphs are generated using random graph technique with two outlier settings. We show the fraction of outliers for each group as different colours in this figure.

Next, we analyze the performance of binary classifiers trained on graph embedding vectors built using different combinations of feature vectors (Table 2).

In Fig. 5 (left), we show the model accuracy for each feature set. Focusing on single feature models (M-0 to M-5), we can see that graph embeddings built on top of **Closeness Centrality** (M-4) perform the best, while models built using **Eigenvector Centrality** (M-5) perform poorly. It is also worth mentioning that **Expansion** (M-0), an easy and fast to compute node feature, does very well. The predictive power of **Closeness Centrality** as a node feature comes from the fact that outlier nodes are, on average, closer to other nodes in the network, since they do not belong to any of the communities but rather randomly connected to the entire network. Therefore, this can be a distinctive factor for graphs with higher number of outlier nodes.

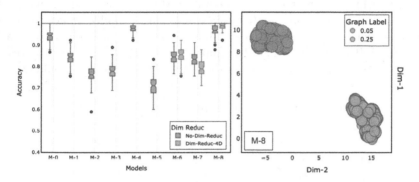

Fig. 5. Left: Accuracy of binary-classifiers built for models M-0 to M-8. Right: two dimensional representation of graph embedding vectors built using features in M-8.

Lastly, we consider composite models, where embeddings are generated from a combination of feature vectors. In Fig. 5 (left), models M-6, M-7, and M-8 are composite models (as defined in Table 2). For each one these models, we run two sets of experiments. One in which we do not apply any dimensionality reduction to the composite feature vectors, before passing them to the graph embedding layer. And, another where we apply dimensionality reduction using UMAP, to reduce the dimension of the composite feature vector to $k = 4$. We can see that in all three cases, dimensionality reduction does not have a significant effect on the performance of the models. One thing worth highlighting is that the composite model (M-8) built using all features with $k = 4$ performs the best. This hints at the fact that one could capture a wide variety of features in the feature computation layer, then reducing the feature space using a technique such as UMAP to allow for a better performance of a given machine learning model at hand.

3.2 Real-World Networks

In this section, we explore the performance of our framework on a collection of real-world networks. This collection was acquired from the Benchmark Data Set provided by the department of computer science of TU Dortmund[9]. A summary of networks used for our experiments are provided in Table 3. Here, we consider five real-world networks (**IMDB** [25], **MUTAG** [6], **NCI1** [24], **BZR** [21], and **PROTEINS** [4]) with various sizes and source, which have sub-networks that can be categorized into two classes. Therefore, a natural type of analysis would be to investigate the performance of binary-classifiers trained on graph embeddings generated by **NEExT**. In our analysis, we use the performance of publicity available models as a benchmark for comparing **NEExT** to other techniques. In this exploratory stage of our project, the goal is not to seek the best performing model at all cost, but rather provide a framework that can easily create models with reasonable performance compared to state-of-the-art techniques, while keeping model explainability. Moreover, note that some models are trained on additional metadata available for nodes as well es edges. We do not do it at present but it would be easy to incorporate such additional information in our model. It is expected that after appropriate selection of node features and fine-tuning the model, the accuracy of the corresponding models should increase.

Table 3. Summary of Real-World Networks.

Name	# of Graphs	# of Classes	Avg. # of Nodes/Edges
IMDB	1000	2	19.77/96.53
MUTAG	188	2	17.93/19.79
NCI1	4110	2	29.87/32.30
BZR	405	2	35.75/38.36
PROTEINS	1113	2	39.06/72.82

In our experiments on the real-networks listed in Table 3, we build $d = 24$ dimensional graph embeddings using the approximate Wasserstein technique on top of $k = 24$ dimensional feature vectors computed on each sub-network. Here, we use 4-dimensional concatenated **LSME**, **Expansion**, **Degree Centrality**, **Closeness Centrality**, **Load Centrality**, and **Eigenvector Centrality** as our feature vectors. We then train the XGBoost binary classifier on top of the $d = 24$ dimensional graph embedding feature vectors. Similarly to the approach taken before, we keep a 70/30 train/test split, and repeat each experiment 100 times to build the statistics for our model performance. Lastly, we point out that we do not balance the datasets in our models, to allow the classifiers to capture statistical imbalances in the underlying data distribution.

In Fig. 6 and Table 4, we show the performance of classifiers trained using **NEExT** and benchmark models collected from leaderboard chart available online[10]. Note that the **LB** accuracies are shown as the range of accuracies from various models submited for each dataset. We see that the accuracy of models

---

[9] https://ls11-www.cs.tu-dortmund.de/staff/morris/graphkerneldatasets.

[10] https://paperswithcode.com/task/graph-classification.

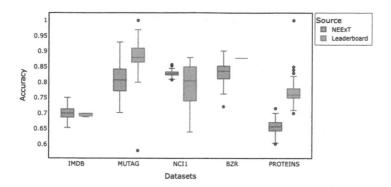

Fig. 6. Accuracy of models built using **NEExT** framework (blue) and publicly available models (red), for various real-world networks. The performance of other models is collected from leaderboard chart available on-line.

Table 4. Summary of Real-World Networks Classification Results. Here, LB Accuracy refers to the range of accuracy of the models from the leaderboard.

Name	Accuracy	Precision	Recall	F1-Score	LB Accuracy
IMDB	0.70 ± 0.02	0.71 ± 0.03	0.69 ± 0.04	0.69 ± 0.02	0.52-0.96
MUTAG	0.81 ± 0.05	0.70 ± 0.08	0.70 ± 0.08	0.69 ± 0.08	0.58-1.00
NCI1	0.83 ± 0.01	0.85 ± 0.02	0.79 ± 0.01	0.82 ± 0.01	0.64-0.88
BZR	0.83 ± 0.03	0.85 ± 0.03	0.95 ± 0.02	0.90 ± 0.02	0.87
PROTEINS	0.66 ± 0.02	0.59 ± 0.05	0.47 ± 0.04	0.52 ± 0.03	0.70-0.85

built using **NEExT** is similar to other models, even without performing any fine-tuning of our models.

4   Conclusion

In this paper, we introduce **NEExT**, the Network Embedding Exploration Tool and show that it can be easily used for feature engineering toward embedding of graphs. This is the beginning of a larger project but the initial experiments we performed so far, some of them reported in this paper, are positive and encouraging to do more work in this space. Here are some natural next steps that we plan to do in the near future.

- Include and test more of our own, carefully designed and explainable, node features. In particular, based on our own personal interests and applications in mind, we plan to add various community-aware node features [14].
- Include more classical node features (such as other centrality measures, degree-degree correlations) and node embeddings (especially structural node embeddings).

- Design and implement an algorithm that automatically selects features and normalizes them in an supervised process, provided that labels for graphs are available.
- Do a grand study comparison with the state-of-the-art methods for graph classification tasks (comparing both the quality of generated embeddings as well as speed). Analyze which types of node features work best for graphs from various domains (explainability).
- Design and implement sampling method which might be useful to embed a large collection of large networks.

References

1. Aggarwal, M., Murty, M.N.: Machine learning in social networks: embedding nodes, edges, communities, and graphs. Springer Nature (2020)
2. Arora, S., Liang, Y., Ma, T.: A simple but tough-to-beat baseline for sentence embeddings. In: International Conference on Learning Representations (2017)
3. Barrett, J., Kamiński, B., Pankratz, B., Prałat, P., Théberge, F.: Self-similarity of communities of the ABCD model. preprint, arXiv (2023)
4. Borgwardt, K.M., Ong, C.S., Schönauer, S., Vishwanathan, S., Smola, A.J., Kriegel, H.P.: Protein function prediction via graph kernels. Bioinformatics **21**(suppl_1), i47–i56 (2005)
5. Cuturi, M.: Sinkhorn distances: Lightspeed computation of optimal transport. In: Burges, C., Bottou, L., Welling, M., Ghahramani, Z., Weinberger, K. (eds.) Advances in Neural Information Processing Systems. vol. 26. Curran Associates, Inc. (2013)
6. Debnath, A.K., Lopez de Compadre, R.L., Debnath, G., Shusterman, A.J., Hansch, C.: Structure-activity relationship of mutagenic aromatic and heteroaromatic nitro compounds. correlation with molecular orbital energies and hydrophobicity. J. Med. Chem. **34**(2), 786–797 (1991)
7. Ester, M., Kriegel, H.P., Sander, J., Xu, X., et al.: A density-based algorithm for discovering clusters in large spatial databases with noise. In: kdd, vol. 96, pp. 226–231 (1996)
8. Kamiński, B., Kraiński, Ł, Prałat, P., Théberge, F.: A multi-purposed unsupervised framework for comparing embeddings of undirected and directed graphs. Network Sci. **10**(4), 323–346 (2022)
9. Kamiński, B., Pankratz, B., Prałat, P., Théberge, F.: Modularity of the ABCD random graph model with community structure. J. Complex Networks **10**(6), cnac050 (2022)
10. Kamiński, B., Prałat, P., Théberge, F.: Artificial benchmark for community detection (ABCD)-fast random graph model with community structure. Network Sci. **9**(2), 153–178 (2021)
11. Kamiński, B., Prałat, P., Théberge, F.: Mining Complex Networks. CRC Press, Boca Raton (2022)
12. Kamiński, B., Prałat, P., Théberge, F.: Artificial benchmark for community detection with outliers (ABCD+o). Appl. Netw. Sci. **8**(1), 25 (2023)
13. Kamiński, B., Prałat, P., Théberge, F.: Hypergraph artificial benchmark for community detection (h–ABCD). J. Complex Networks **11**(4), cnad028 (2023)

14. Kamiński, B., Prałat, P., Théberge, F., Zajac, S.: Predicting properties of nodes via community-aware features. arXiv preprint **2311.04730** (2023). https://doi.org/10.48550/arXiv.2311.04730

15. Kamiński, B., Olczak, T., Pankratz, B., Prałat, P., Théberge, F.: Properties and performance of the ABCDE random graph model with community structure. Big Data Res. **30**, 100348 (2022)

16. Kolouri, S., Naderializadeh, N., Rohde, G.K., Hoffmann, H.: Wasserstein embedding for graph learning. In: International Conference on Learning Representations (2021)

17. Lancichinetti, A., Fortunato, S., Radicchi, F.: Benchmark graphs for testing community detection algorithms. Phys. Rev. E **78**(4), 046110 (2008)

18. Narayanan, A., Chandramohan, M., Chen, L., Liu, Y., Saminathan, S.: subgraph2vec: learning distributed representations of rooted sub-graphs from large graphs. arXiv preprint arXiv:1606.08928 (2016)

19. Narayanan, A., Chandramohan, M., Venkatesan, R., Chen, L., Liu, Y., Jaiswal, S.: graph2vec: learning distributed representations of graphs. arXiv preprint arXiv:1707.05005 (2017)

20. Song, C., Havlin, S., Makse, H.A.: Self-similarity of complex networks. Nature **433**(7024), 392–395 (2005)

21. Sutherland, J.J., O'brien, L.A., Weaver, D.F.: Spline-fitting with a genetic algorithm: a method for developing classification structure- activity relationships. J. Chem. Inf. Comput. Sci. **43**(6), 1906–1915 (2003)

22. Togninalli, M., Ghisu, E., Llinares-López, F., Rieck, B., Borgwardt, K.: Wasserstein weisfeiler-lehman graph kernels. In: Advances in Neural Information Processing Systems, vol. 32 (2019)

23. Villani, C., et al.: Optimal Transport: Old and New, vol. 338. Springer, Berlin (2009). https://doi.org/10.1007/978-3-540-71050-9

24. Wale, N., Watson, I.A., Karypis, G.: Comparison of descriptor spaces for chemical compound retrieval and classification. Knowl. Inf. Syst. **14**, 347–375 (2008)

25. Yanardag, P., Vishwanathan, S.: Deep graph kernels. In: Proceedings of the 21th ACM SIGKDD International Conference on Knowledge Discovery and Data Mining, pp. 1365–1374 (2015)

Efficient Computation of K-Edge Connected Components: An Empirical Analysis

Hanieh Sadri[1]([✉]), Venkatesh Srinivasan[2], and Alex Thomo[1]

[1] University of Victoria, Victoria, BC, Canada
haniehsadri77@gmail.com, thomo@uvic.ca
[2] Santa Clara University, Santa Clara, CA, USA
vsrinivasan4@scu.edu

Abstract. Graphs play a pivotal role in representing complex relationships across various domains, such as social networks and bioinformatics. Key to many applications is the identification of communities or clusters within these graphs, with k-edge connected components emerging as an important method for finding well-connected communities. Although there exist other techniques such as k-plexes, k-cores, and k-trusses, they are known to have some limitations.

This study delves into four existing algorithms designed for computing maximal k-edge connected subgraphs. We conduct a thorough study of these algorithms to understand the strengths and weaknesses of each algorithm in detail and propose algorithmic refinements to optimize their performance. We provide a careful implementation of each of these algorithms, using which we analyze and compare their performance on graphs of varying sizes. Our work is the first to provide such a direct experimental comparison of these four methods. Finally, we also address an incorrect claim made in the literature about one of these algorithms.

Keywords: social networks · community detection · graph algorithms · k-edge connected components · empirical analysis

1 Introduction

Graphs have become increasingly important in today's world due to their ability to capture complex relationships and provide valuable insights [19]. In practical scenarios, we often encounter various data and their relationships which can be effectively depicted using graphs. For instance, they are used in areas like social networks [16], web searches [20], biochemistry [12], biology [2], and road network mapping [7]. Given their widespread use and significance, a lot of research is being conducted to analyze graph data [15].

Identifying communities within graphs, which are essentially clusters of densely connected vertices, is a vital concept due to its wide-ranging applications [10]. In social networking platforms, community detection can be leveraged for

M. Dewar et al. (Eds.): WAW 2024, LNCS 14671, pp. 80–96, 2024.
https://doi.org/10.1007/978-3-031-59205-8_6

friend recommendations, targeted social campaigns, and advertising [28]. Within protein-protein interaction networks, community detection can be applied to recognizing proteins with similar functionality [22]. Meanwhile, in a web-link-based graph, a community might represent a set of web pages sharing substantial commonality, aiding in finding similarities among them [3].

One prevalent method for identifying such clusters and communities is by computing k-edge connected components (cf. [1,5,6,23,29]). These components are induced maximal subgraphs that remain connected after removing $k - 1$ edges. While there are alternative notions of graph density, such as k-core (cf. [9,14,21]) or k-truss (cf. [8,25,27]) components, k-edge connected components often give well-connected communities that k-core and k-truss fail to discover (cf. [1] for more details).

Our work[1] provides a detailed exploration of four main algorithms for computing k-edge connected components. We implement each of these algorithms, compare their performance, and suggest optimization strategies when applicable.

The first algorithm we consider, presented in [5], is based on graph decomposition. Its main idea is to decompose the graph until all the remaining connected subgraphs are k-edge connected. Another algorithm that we explore, introduced in [1], is based on contracting random edges. The idea behind this method is repeatedly finding cuts with sizes less than k and dividing the graph along these cuts. If we reach the point that each connected component has no cut with a size less than k, then they are k-edge connected.

Subsequently, we investigate another algorithm called the early merge and split method, presented in [23], which proposes an improvement to the decomposition algorithm of [5]. Its core idea is to combine vertices that meet the k-connectivity requirement. Lastly, we study an algorithm, given in [6], that computes k-edge connected subgraphs by computing specialized cuts. This algorithm is mainly in the theoretical realm and has resisted implementation until now, which we present in this work.

The main contributions of this paper are summarized as follows:

1. We undertake a thorough implementation and experimental study of the four aforementioned methods for computing k-edge connected components. Our comprehensive analysis assesses each method's strengths, limitations, and applicability. To ensure a fair comparison, we implemented each method in the same programming language and evaluated them under identical environments and setups, utilizing a range of datasets from small to large.
2. An important contribution of our study is our capability to extract the unique features of each method, equipping us to optimize each one effectively. Additionally, we carefully engineer certain algorithms, leading to marked improvements in their scalability.
3. Our study features a direct comparison of the algorithms from [1,5]. Since both were published around the same time, a direct comparison had not been previously conducted, resulting in a gap in comparative analysis. Our

---

[1] The full version of this paper is available here.

work fills this void, providing valuable insights into their relative performance, strengths, and weaknesses.
4. We highlight and correct inaccurate claims made by [23]. Contrary to their claims, we demonstrate that their proposed improvements do not perform as effectively in practice.
5. We provide the first implementation of an algorithm presented in [6] to identify k-edge connected components. This algorithm, due to its abstract nature, had not been previously implemented, even by its original authors. Our implementation sheds light on its practical efficiency and applicability.

2   Definitions

In this section, we begin by introducing some terminology and definitions. Our work deals with undirected and unweighted graphs.

Definition 1. *An* **undirected graph** *G is denoted as $G = (V, E)$, where V is the set of vertices and E is the set of undirected edges. Furthermore, we denote by n and m the sizes of V and E.*

The notions of connectivity of a graph and the degree of a vertex are central to our work.

Definition 2. *Graph G is* **connected** *if for any two vertices $u, v \in V$, there is a sequence of edges $(u, v_1), (v_1, v_2), \ldots, (v_k, v)$ between u and v in G.*

Definition 3. *For a vertex $v \in V$ in $G = (V, E)$, the* **degree** *of v, denoted as $deg(v)$, is defined as the number of edges incident to v.*

Next, we describe the notions of induced subgraphs and cuts in graphs needed to study k-connectedness.

Definition 4. *For a subset $V' \subseteq V$, the subgraph $G[V']$* **induced** *by V' is the subgraph of G with vertex set V' and the edge set $E' \subseteq E$ that only includes the edges from G connecting vertices both in V', i.e. $E' = \{(u, v) \in E \mid u, v \in V'\}$.*

Definition 5. *A* **cut** *in a graph $G = (V, E)$ partitions its vertices into two disjoint subsets S and T. The cut set, denoted $C(S, T)$, consists of all edges with one endpoint in S and the other in T, whose removal disconnects G. We refer to $|C(S, T)|$ as the size of the cut.*

We now formally define the problem we study.

Definition 6. *A graph G is described as* **k-edge connected** *if it continues to be connected even when any $k - 1$ edges are removed. In other words, G is* **k-edge connected** *if the minimum size of a cut in G is at least k. A k-edge connected component of a graph G is an induced subgraph H of G that remains connected even after the removal of any $k - 1$ edges, and there is no larger subgraph in G with this property that contains H.*

3   Related Work

Given a graph G. A k-edge connected component within G is defined as a maximal induced subgraph that retains connectivity despite the removal of any $k-1$ edges. Various methodologies have been developed to identify these components. Decomposition-based algorithms aim to identify k-edge connected components by iteratively decomposing the graph into connected components until each is a k-edge connected component [1,5,23]. On the other hand, cut-based algorithms iteratively cut a graph with connectivity less than k into two parts, continuing this process until the connectivity of all resulting subgraphs is at least k [6,29].

In the realm of graph theory, there is also an alternate definition for k-edge connected components, which we are not examining in this study. According to this other interpretation, a k-edge connected component is recognized as a maximal group V_i of vertices within graph G, where each vertex pair is k-connected in the entire graph G. However, this does not guarantee the same connectivity within the induced subgraph created by V_i, which could be disconnected. Identifying all such maximal vertex groups presents a unique challenge, different from identifying the k-edge connected components of G, which is our research focus. Several studies have explored this alternative concept [11,17,18,24].

The challenge of identifying all k-edge connected components involves determining the k-edge connected components of graph G for all possible values of k. The methodology suggested in [4] revolves around the construction of a hierarchy tree for G, which effectively finds the k-edge connected components for all possible k values. The work in [26] identifies all k-edge connected components, using a definition that targets the discovery of maximal vertex subsets wherein each pair of vertices is k-edge connected within graph G. Finding all k-edge connected components is beyond the scope of this study, and we concentrate on identifying k-edge connected components for a specific k.

4   Algorithms

In this section, we present a thorough analysis of algorithms used to identify k-edge connected components in graphs. Moreover, we shed light on specific enhancements we have implemented to boost the performance of some of these algorithms. Among the studied methods, the graph decomposition algorithm emerged as the most effective for determining k-edge connected components.

4.1   Graph Decomposition Algorithm

An important algorithm, which we refer to as Graph Decomposition (GD), for identifying k-edge connected components was introduced by Chang et al. in [5]. The core idea of the algorithm involves iteratively decomposing graph G. During each iteration, the algorithm identifies subgraphs that are not k-edge connected and further decomposes them. This process continues until all resulting subgraphs are k-edge connected. The outcome is a list of k-edge connected components. High-level pseudocode for this algorithm is provided in Algorithm 1.

Algorithm 1. Computing k-Edge Connected Components
***
1: **Input:** A graph $G = (V, E)$ and an integer k.
2: **Output:** k-edge connected components of G.
3: Initialize a queue Q_g with the graph G as its sole member.
4: **for** each subgraph g in Q_g **do**
5:     $\phi_k(g) \leftarrow$ Decompose(g, k);
6:     **if** $\phi_k(g)$ consists solely of one subgraph **then**
7:         Output $\phi_k(g)$ as a k-edge connected component;
8:     **else**
9:         Enqueue all subgraphs of $\phi_k(g)$ into Q_g;

A key part of the decomposition involves repeatedly applying merge and split operations on what is called the partitioned graph. The idea of a partitioned graph PG is taking a graph G, consisting of vertices V and edges E, and appending additional information to each vertex from a domain D. This is done to track modifications in the graph after vertex merging. In an ordinary graph, vertices do not have extra associated information, thus $D(u) = \{v\}$ for each v in V. However, upon applying a merge operation on vertices v_i and v_j these vertices combine into a single super-vertex, u. Specifically, u is added to PG such that $D(u) = D(v_i) \cup D(v_j)$. Then, edges (u, x) are added to PG if x belongs to the neighbor set of either v_i or v_j. If a vertex x is adjacent to both v_i and v_j, in the resulting partition graph a parallel edge from u to x is created. Following this, vertices v_i, v_j, and their linked edges are removed from PG.

An important part of the GD algorithm is creating several connected subgraphs by removing the edges in a cut of G with a value less than k. This relies on the **Maximum Adjacency Search (MAS)** procedure to compute the minimum cut between a pair of vertices. MAS organizes all vertices of G into a list L. Assume the last vertex in this list is t, and the vertex immediately preceding it is s. In this configuration, the edges adjacent to t in G constitute the minimum s-t cut.

The construction of the list L initiates by randomly selecting a vertex from V and adding it to L. As long as there are remaining vertices not yet in L, a vertex $u \notin L$ is selected. This vertex u is the one with the maximum number of connections to L, mathematically represented as $u = \arg\max_{v \in V \setminus L} w(L, v)$, where $w(L, v)$ denotes the number of edges connecting v with vertices in L. The selected vertex u is then appended to the end of L.

To optimize the identification of the most tightly connected vertex in **MAS**, a specialized data structure is introduced. Let $key(v)$ represent the key of vertex v during the execution of **MAS**, where $key(v) = w(L, v)$, indicating the number of edges between v and vertices in L.

In this data structure, doubly linked lists are employed alongside a head array. The head array specifically holds the first vertex associated with each key value x; to clarify, Head$[x] = v$, where $v \in V$, signifies that v is the head vertex in the doubly linked list with a key value of x. Within this linked list arrangement, three arrays each with a size of $|V|$ maintain the key, and "next", and "previous" indexes (pointers) for each vertex in the list.

Example 1. Refer to Fig. 1 for an illustrative example that visualizes the data structure. From this figure, we observe that $Head(x) = v_i$. To identify the subsequent vertex with a key equal to x, we can use $next(v_i)$. Here we have that $next(v_i) = v_j$ and v_j also has a key of x. Furthermore, we see $previous(v_j) = v_i$.

In addition to the head array and doubly-linked list, we also maintain a value p_0, which represents the current maximum key value among all vertices in the data structure. This value is initialized to 0 and is updated whenever a new vertex is inserted into the data structure.

To update the key of vertex v from x to y, we first remove v from the doubly-linked list represented by $Head(x)$, and then insert v into the doubly-linked list $Head(y)$. Additionally, `max-key` is updated to y if $y > p_0$.

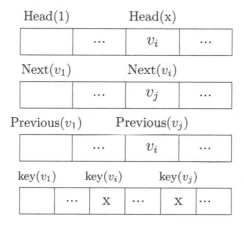

Fig. 1. Data structure

To extract the next vertex with the largest key value, we first decrement p_0 until $Head(p_0)$ is not null. We then report the first vertex pointed to by $Head(p_0)$ and remove that vertex from the doubly-linked list.

An algorithm called `Decompose-LMS` integrates the aforementioned ideas into an efficient method for implementing decomposition. In this algorithm, a method called `MAS-LMS` is iteratively applied until the edge set of the graph gets empty. `MAS-LMS` employs linear data structures and incorporates early merge and list-sharing optimizations explained in the following.

During a single iteration of `MAS-LMS`, it is possible to merge multiple pairs of vertices, provided each pair is guaranteed to be k-connected in the input graph. This strategy is known as Early Merge. If `MAS-LMS` identifies a minimum cut smaller than k, the minimum cut produced by the subsequent split operation can be directly obtained from the existing list L. This eliminates the need to execute MAS-LMS on the newly formed graph. The capability to reuse the list L across multiple instances of MAS following split operations is termed List Sharing.

The time complexity of the Decompose-LMS is given by $O(l \cdot |E|)$, where l denotes the number of repetitions of the MAS-LMS procedure. The overall time complexity of GD Algorithm using Decompose-LMS for decomposition is $O(h \cdot l \cdot |E|)$, where h is a parameter representing the number of times Decompose-LMS is called within Algorithm 1. As pointed out by [5], h is typically a small integer for real-world graphs.

4.2   Random Contraction Algorithm

The concept of finding k-edge connected components through random contractions was introduced by Akiba et al. in [1]. The proposed algorithm, we refer to as RC, is a unique application of random contraction. This methodology has historically been a theoretical tool for addressing cut problems, as noted in [13]. The random contraction method of [1] involves selecting edges at random and contracting them until no edges are left in the graph. Contracting an edge involves the removal of the edge and the subsequent merging of its two endpoints.

In this algorithm, the graph is represented by a dictionary of dictionaries. For every vertex v, there's an associated dictionary, h_v; the keys are neighboring vertices, while the values stand for the edge weights.

When an edge (u, v) is contracted, the algorithm merges the dictionaries h_u and h_v. To ensure that this merge is done efficiently, the edges from the smaller dictionary are always inserted into the larger one. So, supposing that h_u is smaller than h_v, the transfer process would involve moving edges from h_u to h_v. During this transfer, a vertex x from h_u to h_v, if h_v doesn't already contain the edge (v, x), then the algorithm simply adds the edge (v, x) to h_v. But, if h_v contains the edge (v, x), the weight of (v, x) in h_v is increased by the weight of (u, x) found in h_u. Additionally, we add v to h_x, with a weight of the edge (v, x).

The next step involves removing u from the overall dictionary. This necessitates navigating through all of u's neighbors, as specified in h_u, and systematically excluding u from their neighbor lists. Once this step is completed, h_u can be safely removed from the graph.

Algorithm 2 for finding connected-subgraph of G by random contraction starts with generating a copy of Graph G, denoted as G'. While G' still contains edges, it randomly selects an edge for contraction. After this contraction, the degree of a vertex, say, u, might decrease and become less than k. If this happens, the subgraph formed by the vertices merged with u is added to the output. Also, all the edges connected to u are removed, and u is excluded from G'. This entire set of actions constitutes one iteration.

The iterations are then repeated for each subgraph induced by a connected component from the preceding iteration, and this is done for a predefined number of times. The precise number of iterations required depends on the graph and is determined by checking that no new output is created in some iteration. Akiba et al. show that the total time complexity of RC algorithm is $O(|E| \log(n))$.

Algorithm 2. Basic Iteration

1: **procedure** CONTRACTANDCUT(G, k)
2:     $G' \leftarrow G$
3:     **while** G' is not empty **do**
4:         **if** exists $u \in V(G')$ such that $d(u) < k$ **then**
5:             $U \leftarrow$ original vertices contracted to u
6:             output $G[U]$
7:             Remove u from G'
8:         **else**
9:             Choose an edge (v, w) in G' at random
10:            Contract v and w in G'

One crucial observation is that while each iteration yields connected components, they may not always be k-edge connected. A component that remains unbroken in a subsequent iteration is indeed k-edge connected.

To streamline the process, [1] introduces a method called **forced contraction**. The underlying principle is simple: if an edge between two vertices, u and v, carries a weight of k or more, it is beneficial to contract these vertices immediately.

The reason behind such immediate contraction lies in the understanding that if the edge's weight is at least k, there's no way to separate the two vertices by a cut smaller than this weight. So contracting these vertices will not ruin any potential cut of size less than k and by contracting them, the chances of finding other cuts smaller than k increase.

The algorithm doesn't specify the number of iterations, which can lead to potential errors. However, it is suggested in [1] that by setting the number of iterations to $O(\log_2 n)$, the error probability can be reduced to as low as $\frac{1}{1000}$. In our tests, we carefully chose the number of iterations to accurately determine maximal-k-edge connected components.

Our contribution in this section is an optimization of the random contraction implementation, drawing inspiration from a graph representation technique presented in [5]. This array-based method provides an advantage during graph traversal: all the graph data resides in a contiguous block of memory. This arrangement accelerates the process compared to traditional adjacency lists. The RC algorithm, utilizing a streamlined data structure, is named RCF.

In this structure, a graph is represented using four arrays. Central to this representation is the graph_head array. For every vertex v in V, graph-head(v) indicates the index of the starting neighbor in the value array. The "value" array keeps the actual neighbors of the vertices in the graph. Using the "next" array, we can identify the indices of subsequent neighbors.

When next(i) = -1, it indicates the end of that vertex's adjacency list. The "previous" array functions like next, but instead points to the preceding neighbor. The bidirectional aspect of the graph is addressed by the reverse array. For an edge (u, v) located at index i in value array, "reverse" keeps its counterpart edge (v, u) in a manner that if $value(i) = v$, then $value(reverse(i))$ would represent u. So based on what we explained, to retrieve all neighbors of a vertex v, we start at graph-head(v) and keep iterating using the next array until reaching -1.

When vertices u and v are contracted, we add v's neighbors to u's neighbor list. If a vertex x is a neighbor to both u and v, it will appear twice in u's list after the merge. If we store edge weights in a separate array, updating these during a merge would require checking if a neighbor of v is also a neighbor of u before adjusting the weight. This check takes $O(n)$ time for our structure. As keeping weights implies additional running time when using our proposed data structure, we do not maintain weights in our implementation. As such, we forego forced contractions, which depend on these weights and hence cannot be applied. Empirical testing on diverse datasets reveals that our array-based implementation RCF consistently outperforms the original algorithm based on dictionaries.

4.3 Early Merging and Splitting

The early merging and splitting algorithm, known as the MSK algorithm, was proposed by Sun et al. [23]. It determines the k-edge connected components by sequentially examining an ordered list of vertices. This order is established based on each vertex's connectivity within the graph. As the algorithm processes this list, it merges any two vertices exhibiting k-edge connectivity into a singular super-vertex. Conversely, if the vertex pair does not satisfy the condition of k-edge connectivity (there exists a cut of size less than k separating the two vertices), the edges of the cut are removed from the graph, and the graph is decomposed into two subgraphs. This procedure continues iteratively on the resultant subgraphs until every one of them qualifies as a k-edge connected component.

This method presents notable similarities to the approach of [5] described in Sect. 4.1. Both strategies utilize the MAS procedure to find minimum s-t cuts. Additionally, in either approach, multiple vertex pairs can be merged in one MAS iteration using the early merging technique, as long as the connectivity between s and t remains at least k.

The key distinction lies in how the cuts are split. In Decompose-LMS, cuts are split after completing the list L for the MAS procedure. In contrast, in MSK, before inserting vertices into L during MAS, we check the weight between these vertices and those not in L. If this weight is below k, the graph splits into two subgraphs: one formed by the vertices in L and another from the original vertices that had merged with vertices in $V \setminus L$ in prior iterations. So, in MSK, as soon as the weight between vertices in L and vertices in $V \setminus L$ is less than k, we immediately split the cuts and decompose the graph into two subgraphs. The time complexity of the MSK is $O(r \cdot m)$ where r indicates the number of iterations in the MSK algorithm and m is the number of edges in the graph.

Incorrect claim. In the MSK paper, the authors view the algorithm of [5], as detailed in the graph decomposition section, as an approximation rather than an exact method for identifying k-edge connected components. This interpretation stems from the fact that, during the `Decompose-LMS` algorithm, there exists a possibility for two vertices to merge provided their connectivity is at least k in the input graph. Yet, when certain cuts are removed in subsequent iterations, the connectivity between these vertices might drop below k.

The authors of [23] incorrectly deduced that the results from a singular `Decompose-LMS` invocation amounted to the k-edge connected component for the graph decomposition procedure as illustrated in Sect. 4.1. This misinterpretation underpins their labeling of the graph decomposition as an approximate algorithm. However, the true k-edge connected components are extracted from Algorithm 1 in Sect. 4.1. Notably, `Decompose-LMS` is recursively executed until all components achieve k-connectivity.

Upon unifying the implementation environments of both algorithms and integrating the heap data structure from `Decompose-LMS` into MSK to optimize the MAS procedure, our comprehensive tests favored graph decomposition over MSK. We delve into the details of these findings in the experiments section.

4.4 Local Cut Detection

Chechik et al. in [6] present yet another algorithm for computing maximal k-edge connected components in directed graphs [6], which we refer to as LCD. A directed graph is k-edge connected if it is strongly connected whenever fewer than k edges are removed. While their algorithm targets directed graphs, it can also be used for undirected graphs by replacing each undirected edge with two bidirectional edges.

The main idea of this method is to identify a small subgraph, with at most $\sqrt{m}$ edges, that's well separated from the rest of the graph. This is done by performing DFS traversals that, when starting from some vertex, traverse mostly the edges within this small subgraph and a few edges outside it. This subgraph is considered "well-separated" because it isn't k-edge connected to the other parts of the graph. After identifying it, the edges connecting this subgraph to the rest of the graph are removed. This step is referred to as `local cut detection`. The `local cut detection` is repeated until no cut in the graph is smaller than k. More specifically, a `local cut detection` identifies a k-edge-out component of a vertex u, which is a subgraph that contains u and has no more than k edges extending from the subgraph to the remainder of the graph. Similarly, a k-edge-in component is computed in the same manner but on the inverse graph.

The algorithm to compute k-edge connected components begins with a given graph and a list L, which initially contains the vertices of the graph. After initializing the list L and setting the initial number of edges in the graph as m, the algorithm checks if the graph has a cut of less than k edges. If not, the graph is returned as a k-edge connected component. Otherwise, a loop starts and continues until L is not equal to the empty set and the graph contains more than $2k\sqrt{m}$ edges. In each iteration, the k-edge-out component and k-edge-in component are calculated for a vertex u extracted from list L. If either of these is not equal to the empty set, the algorithm removes the edges with endpoints in the result of the k-edge-out or k-edge-in and continues. After the loop, the strongly connected components (SCCs) of the resulting graph are calculated.

A set U is initialized as an empty set. For each SCC, the algorithm calculates a $k-1$ cut set if it exists, removes the $k-1$ cut set, and computes the SCCs of the resulting graph. Edges between the SCCs are removed and their endpoints are saved in a list. For each SCC, a list L' is created and populated with the vertices of the SCC that are in the saved list of endpoints. The algorithm then recursively processes the SCC and L', and unites the results with U.

The run-time of this algorithm that computes k-edge connected components using local cut detection is $(O((2k)^{k+1} \times \sqrt{m} \times m \log n)$, where k represents the edge connectivity. In the paper, k is treated as a constant; hence, the factor $(2k)^{k+1}$ is omitted from the analysis. The overall complexity stands at $O(m\sqrt{m} \log n)$ when this factor is disregarded. Nonetheless, in real-world applications, maximum k can be quite large ranging between 30 to 80 for actual graphs. For such magnitudes, the algorithm presented in [6] becomes impractical. For instance, in the soc-epinions graph used in our experiments, the maximum k value is 67. This translates to $(2k)^{2k+1}$ being roughly 290,000,000. As a result, computing the 67-connected components for this graph using the current algorithm is impractical for real-world scenarios.

5   Experiments

Our study conducts a comparative assessment of selected algorithms, emphasizing their running times as the primary metric. We analyzed the efficiency of each algorithm under standardized conditions. All algorithms were implemented in Java to ensure consistency and the source code can be accessed at GitHub[2]. We performed our experiments on Compute Canada's Cedar5, a high-performance computing cluster equipped with dual 6-core 2.10 GHz Intel Xeon CPUs. Each test was allocated 32GB RAM.

We evaluated our algorithms on eight real graphs. Real graphs, derived from actual data sources, contain inherent structures and patterns that represent real-world scenarios. All the graphs are obtained from the Stanford SNAP library2, and detailed descriptions of these graphs can also be found there. The sizes of these graphs are shown in Table 1. By varying the sizes of these graphs, we can evaluate our algorithms across different graph structures, from small to large.

5.1   Small Graphs

The small datasets that we considered are bird with 958 edges, feather-lastfm-social with 27,806 edges, ego-Facebook with 88,234, and feather-deezer-social with 92,752. We plot the run times achieved by RC, RCF, GD, and MSK on all the small datasets. However, for LCD, we only recorded its runtime for the bird dataset. This algorithm was not scalable for the other datasets; it failed to produce results even after an extended period of 48 h.

---

[2] https://github.com/Haniehsadri/KECC.

Table 1. Datasets

Data Set	# Vertices	# Edges
bird	129	954
feather-lastfm-social	7,624	27,806
ego-Facebook	4,039	88,234
feather-deezer-social	28,281	92,752
musae-git	37,700	289,003
soc-epinions	75,879	508,837
com-DBLP	317,080	1,049,866
amazon	262,111	1,234,877

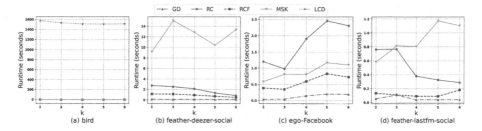

Fig. 2. Performance Analysis on Small Datasets: The figure compares the runtime of various algorithms on small datasets. It underscores the notable efficiency of GD and RCF while illustrating the pronounced slower runtime of LCD.

From Fig. 2, it is evident that the LCD is significantly slower than the other algorithms. While most algorithms process the bird dataset in a negligible time (close to 0 s), on average, the LCD algorithm takes 1500 s to process this dataset. From Figs. 2.b, 2.c, and 2.d, it is clear that GD requires less time to complete and outperforms the other algorithms in all instances. RCF is faster than both the RC and MSK. When comparing RC to MSK for smaller data sets, there are situations where RC performs better, and in other instances, the MSK demonstrates a faster runtime. From all the plots in 2, it can be seen that the GD outperforms other algorithms for small datasets.

5.2   Medium and Large Graphs

For medium and large graphs, we considered several datasets: the musae-github graph with 289,003 edges, soc-epinions with 508,837 edges, com-DBLP which contains 1,049,866 edges, and the sizable amazon0302 graph which comprises 1,234,877 edges. We plotted the run times achieved by the four algorithms and the results can be observed in Fig. 3 Table 2.

In evaluations involving medium and large datasets, the GD algorithm consistently outperforms the other algorithms, much like its performance with smaller

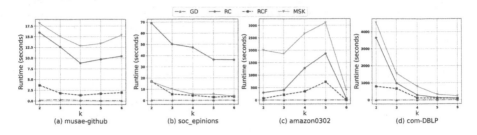

Fig. 3. Performance Analysis on Medium to Large Datasets: The figure shows runtime performance for four algorithms on different graph sizes. It highlights GD's steady efficiency and RCF's advantages over RC. For the largest dataset, amazon, runtime rises with k until $k = 5$ and then sharply drops at $k = 6$. This trend is due to the increase in k-connected components up to $k = 5$, followed by a significant reduction at $k = 6$, as detailed in .

datasets. The RCF algorithm outperforms both the MSK and the RC algorithms in terms of efficiency. For medium-sized graphs, MSK yields better results when compared to RC. However, for larger graphs, MSK starts to falter in terms of scalability, a trend that's evident in Figs. 3.c and 3.d. We mention here that we are the first to be able to run our MSK implementation to large datasets. The original authors were only able to run up to the Soc-Epinions dataset.

A similar trend observed across plots in Fig. 3 is that, for most cases, the processing times of all four algorithms decrease as k increases. As k becomes larger, the graph, after removing all vertices with degrees less than k, becomes smaller, leading to faster execution of all algorithms. However, this trend might vary for certain graphs. For instance, with the amazon dataset (shown in Table 2), as k increases, the number of components also increases significantly, necessitating a considerable increase in the number of iterations for RC. It is only when k goes from 5 to 6 that the number of components drops drastically from 2653 to 32. At this point, the running time decreases significantly.

Another analysis we perform is the evaluation of the performance of each algorithm on datasets of varying sizes, assessing how their efficiency changes as the data size transitions from small to medium, and from medium to large (see Fig. 4). We demonstrated the experiment with $k = 2$. However, based on our experiments with other possible values of k, the results followed the same pattern as with $k = 2$.

Starting with the GD algorithm, our experiments show that it performs efficiently and consistently across graphs of different sizes. Even when the graph size increases, the increase in time is not significant. As can be seen in Fig. 4, the running time of GD remains consistently low whether the data size changes from small to medium or from medium to large.

Moving on to the RC algorithm, it performs well for small to medium-sized graphs. However, its efficiency decreases for larger graphs, such as the amazon dataset. Figure 4 shows a noticeable increase in the running time when transitioning from medium to large-sized graphs. The RCF algorithm performs well for both small and medium datasets. The efficiency of this algorithm slightly changes when the data size increases from medium to large.

Examining the MSK algorithm, it operates efficiently for small to medium datasets. Yet, for larger datasets, such as amazon, its performance significantly drops. There's a distinct increase in its running time when the dataset size shifts from medium to large.

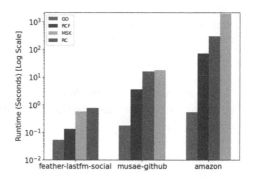

Fig. 4. Comparing running times for GD, RC, RCF, and MSK on three datasets with different sizes for $k = 2$. The figure underscores the consistent efficiency of GD, the waning performance of the RC and MSK on larger datasets, and the subtle transition in the efficiency of RCF from medium to large datasets. A similar trend was observed for $k > 2$ (not shown here).

5.3   Evaluation of Optimization Techniques for RC

In this subsection, we evaluate the effectiveness of the optimization techniques introduced in Sect. 4.2 for the RC algorithm. The experiments conducted on various datasets underscore that the RCF method consistently outperforms the original RC approach in empirical tests even without implementing forced contraction. To elucidate this distinction further, we examine both algorithms on the amazon dataset.

We emphasize that the components produced in each iteration of RC and RCF are connected, but may not always be k-edge connected. Therefore, multiple iterations with each algorithm are required to achieve the desired number of exact k-edge connected components. We recall that both RC and RCF are randomized algorithms, carrying a small (but non-zero) probability of failing to identify all the k-ecc components. We used the results from GD as a benchmark to determine the number of iterations needed for RC and RCF to identify these components.

In the plots illustrated in Fig. 2 and 3, we executed both the RC and RCF algorithms using the appropriate number of iterations. From the plots, it is evident that even though the RC needs fewer iterations to reach the desired number of components, RCF consistently outperforms it time-wise. As an example, let us consider the amazon dataset from Table 2, with $k = 4$. The RC algorithm identifies the 1,287 4-edge connected components in just 7 iterations. In contrast, the

Table 2. Number of extracted k-ecc's after each iteration for RC and RCF on amazon. Also shown are the numbers of k-ecc's obtained by GD, which serve as ground truth numbers. Recall that RC and RCF are randomized algorithms with some small (but non-zero) probability of not being able to discover all the k-ecc's. Also, RCF forgoes the forced random contractions that RC does.

	k=2			k=3				k=4			k=5				k=6
Iter	1	3	6	1	5	7	11	1	7	14	1	5	14	16	1
RC	355	367	367	681	756	777	777	876	1287	1287	5	2437	2653	2653	32
RCF	350	363	367	402	741	763	777	678	1202	1287	5	2326	2337	2653	32
GD	367			777				1287			2653				32

RCF algorithm needs 14 iterations to compute them. Yet, the runtime of RCF, even with 14 iterations, stands at 346 s, whereas the RC, with its 7 iterations, takes a significantly longer 1,277 s.

6   Discussion

In theory, all the algorithms presented in Sect. 4 possess linear time, or near linear, in m, complexity. However, our experiments show variations in their performance across different datasets. From our results, the GD algorithm is the most efficient in determining k-edge connected components. On the other hand, the RC algorithm struggles, especially with larger graph sizes. Notably, after improving its data structures and obtaining RCF, its speed improved considerably, making it more suitable for larger networks. However, even after these improvements, the optimized version, while better than its predecessor, still lags behind the GD algorithm.

The MSK algorithm was further refined by incorporating the heap data structure from GD, reducing its time complexity from $O(m + n \cdot \log n)$ to $O(m \cdot r)$. However, even after these changes, it is outperformed by the GD and RCF algorithms for larger datasets. For smaller and medium-sized datasets, it performs better than the RC algorithm. But for larger networks, its efficiency decreases, and RC becomes more efficient.

Regarding the LCD algorithm its practical scalability is limited, deeming it suitable only for small datasets. The original researchers only assessed its time complexity without any empirical tests. Its practical application appears limited, demonstrating feasibility only for very small networks.

References

1. Akiba, T., Iwata, Y., Yoshida, Y.: Linear-time enumeration of maximal k-edge-connected subgraphs in large networks by random contraction. In: CIKMm, pp. 909–918 (2013)
2. Barabási, A.L., Oltvai, Z.N.: Network biology: understanding the cell's functional organization. Nat. Rev. Genet. **5**(2), 101–113 (2004)

3. Brin, S., Page, L.: Anatomy of a large-scale hypertextual web search engine. Comput. Netw. ISDN Syst. **30**(1–7), 107–117 (1998)
4. Chang, L., Wang, Z.: A near-optimal approach to edge connectivity-based hierarchical graph decomposition. VLDB J. **15**, 1–23 (2023)
5. Chang, L., Yu, J.X., Qin, L., Lin, X., Liu, C., Liang, W.: Efficiently computing k-edge connected components via graph decomposition. In: SIGMOD, pp. 205–216 (2013)
6. Chechik, S., Hansen, T.D., Italiano, G.F., Loitzenbauer, V., Parotsidis, N.: Faster algorithms for computing maximal 2-connected subgraphs in sparse directed graphs. In: SODA, pp. 1900–1918 (2017)
7. Delling, D., Sanders, P., Schultes, D., Wagner, D.: Engineering route planning algorithms. In: Lerner, J., Wagner, D., Zweig, K.A. (eds.) Algorithmics of Large and Complex Networks. LNCS, vol. 5515, pp. 117–139. Springer, Heidelberg (2009). https://doi.org/10.1007/978-3-642-02094-0_7
8. Esfahani, F., Daneshmand, M., Srinivasan, V., Thomo, A., Wu, K.: Scalable probabilistic truss decomposition using central limit theorem and h-index. Distrib. Parallel Databases **40**(2–3), 299–333 (2022)
9. Esfahani, F., Srinivasan, V., Thomo, A., Wu, K.: Efficient computation of probabilistic core decomposition at web-scale. In: EDBT, pp. 325–336 (2019)
10. Fortunato, S.: Community detection in graphs. Phys. Rep. **486**(3–5), 75–174 (2010)
11. Galil, Z., Italiano, G.F.: Reducing edge connectivity to vertex connectivity. ACM SIGACT News **22**(1), 57–61 (1991)
12. Jeong, H., Mason, S.P., Barabási, A.L., Oltvai, Z.N.: Lethality and centrality in protein networks. Nature **411**(6833), 41–42 (2001)
13. Karger, D.R.: Global min-cuts in RNC, and other ramifications of a simple min-cut algorithm. In: SODA, pp. 21–30 (1993)
14. Khaouid, W., Barsky, M., Venkatesh, S., Thomo, A.: K-core decomposition of large networks on a single PC. PVLDB **9**(1), 13–23 (2015)
15. Leskovec, J., Chakrabarti, D., Kleinberg, J., Faloutsos, C., Ghahramani, Z.: Kronecker graphs: an approach to modeling networks. J. Mach. Learn. Res. **11**, 985–1042 (2010)
16. Mislove, A., Marcon, M., Gummadi, K.P., Druschel, P., Bhattacharjee, B.: Measurement and analysis of online social networks. In: IMC 2007, pp. 29–42 (2007)
17. Nagamochi, H., Ibaraki, T.: A linear time algorithm for computing 3-edge-connected components in a multigraph. Jpn. J. Ind. Appl. Math. **9**, 163–180 (1992)
18. Nagamochi, H., Watanabe, T.: Computing k-edge-connected components of a multigraph. IEICE Trans. Fundam. Electron. Commun. Comput. Sci. **76**(4), 513–517 (1993)
19. Newman, M.: The structure and function of complex networks. SIAM Rev. **45**(2), 167–256 (2003)
20. Page, L., Brin, S., Motwani, R., Winograd, T.: The PageRank citation ranking: Bringing order to the web. In: Stanford InfoLab (1999)
21. Seidman, S.: Network structure and minimum degree. Soc. Netw. **5**(3), 269–287 (1983)
22. Spirin, V., Mirny, L.A.: Protein complexes and functional modules in molecular networks. Proc. Natl. Acad. Sci. **100**(21), 12123–12128 (2003)
23. Sun, H., et al.: Efficient k-edge connected component detection through an early merging and splitting strategy. Knowl. Based Syst. **111**, 63–72 (2016)
24. Tsin, Y.H.: Yet another optimal algorithm for 3-edge-connectivity. J. Discr. Algorithms **7**(1), 130–146 (2009)

25. Wang, J., Cheng, J.: Truss decomposition in massive networks. PVLDB 5(9), 812–823 (2012)
26. Wang, T., Zhang, Y., Chin, F.Y., Ting, H.F., Tsin, Y.H., Poon, S.H.: A simple algorithm for finding all k-edge-connected components. PLoS ONE 10(9), e0136264 (2015)
27. Wu, J., Goshulak, A., Srinivasan, V., Thomo, A.: K-truss decomposition of large networks on a single consumer-grade machine. In: ASONAM, pp. 873–880 (2018)
28. Yang, J., Leskovec, J.: Defining and evaluating network communities based on ground-truth. Knowl. Inf. Syst. 42(1), 181–213 (2015)
29. Zhou, R., Liu, C., Yu, J.X., Liang, W., Chen, B., Li, J.: Finding maximal k-edge-connected subgraphs from a large graph. In: Proceedings of the 15th International Conference on Extending Database Technology, pp. 480–491 (2012)

The Directed Age-Dependent Random Connection Model with Arc Reciprocity

Lukas Lüchtrath[1]($\boxtimes$) and Christian Mönch[2]

[1] Weierstrass Institute for Applied Analysis and Stochastics, Berlin, Germany
lukas.luechtrath@wias-berlin.de
[2] Johannes Gutenberg Universität Mainz, Mainz, Germany
cmoench@uni-mainz.de

Abstract. We introduce a directed spatial random graph model aimed at modelling certain aspects of social media networks. We provide two variants of the model: an infinite version and an increasing sequence of finite graphs that locally converge to the infinite model. Both variants have in common that each vertex is placed into Euclidean space and carries a birth time. Given locations and birth times of two vertices, an arc is formed from younger to older vertex with a probability depending on both birth times and the spatial distance of the vertices. If such an arc is formed, a reverse arc is formed with probability depending on the ratio of the endpoints' birth times. Aside from the local limit result connecting the models, we investigate degree distributions, two different clustering metrics and directed percolation.

Keywords: directed random geometric graph · social network · percolation · clustering metrics

1 Motivation and Background

A well established paradigm for the modelling of large, sparse networks occurring in the context of the world wide web or social media is the preferential attachment (PA) mechanism [3,26]: networks are built recursively by adding nodes and links in such a way that new nodes prefer to be connected to existing nodes if they have a high degree. PA networks typically are *scale-free* [8], i.e. have power law degree distributions. They also tend to be *robust under random attack* if the degree distribution is sufficiently heavy-tailed [7]. If their nodes are embedded into space and the attachment mechanism interacts with the spatial geometry, then they can further be shown to display *clustering* effects [29]. However, an obvious but important fact that was discussed very early in the context

CM's research is supported by the German Research Foundation (DFG) Priority Programme 2265, grant no. 4439160. LL gratefully received support by the Leibniz Association within the Leibniz Junior Research Group on *Probabilistic Methods for Dynamic Communication Networks* as part of the Leibniz Competition (grant no. J105/2020).

M. Dewar et al. (Eds.): WAW 2024, LNCS 14671, pp. 97–114, 2024.
https://doi.org/10.1007/978-3-031-59205-8_7

of webgraph modelling, see e.g. [10], is that the link structure of typical real world networks is intrinsically directed. Therefore such networks ought to be represented as directed graphs (*digraphs*). The currently available mathematical literature on scale-free digraph models is surprisingly sparse. There are essentially only two models for which rigorous mathematical results answering most of the basic questions of interest for network science are available: inhomogeneous random digraphs [6,12] and the directed configuration model [13,17,34]. Most preferential attachment models are intrinsically directed as well, albeit in a deterministic way as an artefact of the recursive modelling scheme. Using solely the 'arrow of time' to direct a preferential attachment network is in general a poor modelling choice, unless some recursive effect dominates the real world network one intends to model[1]. Therefore PA networks are typically studied as undirected graphs. An exception is the recent paper [14] in which a non-spatial preferential attachment model with *arc reciprocity* is introduced.

Here, we propose a *spatial* PA-style directed network with arc reciprocity. One can think of the construction mechanism as a recursive attachment scheme in which each preferentially established arc triggers the potential creation of the corresponding reverse arc. Because degree-based PA mechanisms lead to relatively complicated dependencies among the edges, we build our model from the simplified *age-based* PA-scheme introduced in [21]. The approach of translating degree into age [19] relies on the intuitive fact that preferential attachment creates a strong positive correlation between age and high degree: only vertices that have been in the system for a long time acquire a sufficiently high degree to attract many new connections. Conversely, the first few vertices of the PA-scheme tend to have very high degrees [8,20].

We provide two variants of the model in the following section, an infinite version that is analytically easier to treat and a finite version which is more appealing from a modelling point of view and related to the infinite model by a local limit theorem. Section 3 then investigates local properties of the model such as degree distribution and clustering metrics and Sect. 4 is devoted to the existence of large weakly connected components.

2   Model Introduction

2.1   The Directed Age-Dependent Random Connection Model

We first describe our model ad hoc as an infinite digraph. In Sect. 2.2 we detail how this digraph arises as the weak local limit [2,4] of a directed PA-type sequence of growing networks. The vertex set of the *directed age-dependent random connection model* (DARCM) is given by a unit intensity Poisson process $\mathcal{X}$ on $\mathbb{R}^d \times (0,1)$. We denote the vertices by $\mathbf{x} = (x, t_x) \in \mathcal{X}$ and call $x \in \mathbb{R}^d$ the vertex' *location* and $t_x \in (0,1)$ the vertex' *birth time*. For two vertices $\mathbf{x} = (x, t_x)$ and $\mathbf{y} = (y, t_x)$ with $t_y < t_x$, we refer to $\mathbf{y}$ as being *older* than $\mathbf{x}$ and to $\mathbf{x}$ as being

---

[1] One example that comes to mind are scientific citation networks in which a publication sends an arc to every publication it references.

younger than **y**, respectively. Almost surely, there are no vertices born at the same time. Our choice of identifying the second vertex coordinate as birth time is rooted in the local limit representation of Sect. 2.2. To define the distribution of arcs in the graph we introduce the following parameters:

(i) the *spatial decay exponent* $\delta > 1$, defining the *spatial profile* $\rho(x) = 1 \wedge x^{-\delta}$ for $x > 0$;

(ii) the *power-law parameter* $\gamma \in (0,1)$ tuning the tail decay of the degree distribution;

(iii) the *edge intensity* $\beta > 0$;

(iv) the *reciprocity exponent* $\Gamma > 0$.

The digraph $\mathscr{D} = \mathscr{D}[\beta, \gamma, \delta, \Gamma]$ is built using these parameters via the following procedure:

(A) Given $\mathcal{X}$, each vertex $\mathbf{x} = (x, t_x)$ forms an arc to each vertex $\mathbf{y} = (y, t_y)$ with $t_y < t_x$ independently of all other potential arcs with probability

$$\rho\big(\beta^{-1} t_y^{\gamma} t_x^{1-\gamma} |x - y|^d\big). \tag{1}$$

If an arc is formed during this step, we denote this by $\mathbf{x} \to \mathbf{y}$ or $\mathbf{y} \leftarrow \mathbf{x}$.

(B) Given $\mathcal{X}$ and all arcs $\mathbf{x} \to \mathbf{y}$ created in (A), each vertex $\mathbf{y} = (y, t_y)$ sends a *reverse arc* to each $\mathbf{x} = (x, t_x)$ with $\mathbf{x} \to \mathbf{y}$ independently of all other potential reverse arcs with probability $(t_y/t_x)^{\Gamma}$. If such a reciprocal connection is made, we denote this event by $\mathbf{x} \leftrightarrow \mathbf{y}$.

Observe that δ controls the spatial decay of connection probabilities such that vertices in mutually distant locations have a low probability of connecting. This effects gets stronger the larger the value of δ. The strongest spatial restrictions are modelled by the case $\delta \to \infty$ which we identify with the indicator function $\rho = \mathbb{1}_{[0,1]}$. In contrast, the likelihood of a reciprocal connection does not depend on the spatial locations and takes only birth times and the reciprocity exponent into account. The assumptions $\delta > 1$ and $\gamma < 1$ ensure that the expected number of arcs incident to any vertex remains bounded.

For the choice of $\Gamma = 0$ each edge is oriented into both directions, thus $\Gamma = 0$ corresponds to constructing an undirected graph denoted by $\mathscr{G} = \mathscr{D}[\beta, \gamma, \delta, 0]$. The graph $\mathscr{G}$ is the *age-dependent random connection model* introduced in [21]. Clearly, $\mathscr{D}$ can be generated from $\mathscr{G}$ by first pointing all edges in $\mathscr{G}$ from younger end-vertex to older end-vertex and then adding reverse arcs by way of (B).

Let us very briefly motivate our construction. The locations of the vertices describe intrinsic connection affinities and two vertices are more likely to connect if they are spatially close to each other. The age of a vertex indirectly models its performative attractiveness in the graph, the older a vertex the more arcs it attracts. These two sources of connections are intended to model the formation of social media networks: users tend to follow either a friend, i.e. someone with intrinsic affinity to them, or 'influencers', i.e. users who have accumulated a lot of followers already, this effect will become even more apparent in the finite domain version of the model described in the next subsection.

2.2 A Generative Model on Finite Domains and a Local Limit Procedure

We modify our construction to obtain a generative version of the model on a finite domain with periodic boundary conditions, akin to the age-based preferential attachment model of [21]. Chose $\beta > 0$, $\gamma \in (0,1)$, $\delta > 1$, and $\Gamma \geq 0$ to build a growing sequence of directed graphs $(\mathscr{D}_t : t \geq 0)$ in continuous time as follows: At time $t = 0$, the graph $\mathscr{D}_0$ is the empty graph consisting of neither vertices nor edges. Then

- vertices are born successively after independent standard exponential waiting times and are placed uniformly at random and independent of all other locations on the d-dimensional unit torus $[-1/2, 1/2)^d$.
- Let τ_n be the birth time of the n-th vertex placed a location x_n, which we denote by $\mathbf{x}_n = (x_n, \tau_n)$. Given the graph $\mathscr{D}_{\tau_n-}$, the graph build up to time τ_n consisting of the $n-1$ previously arrived vertices, the new vertex x_n forms an arc to each already existing vertex $\mathbf{x}_j = (x_j, \tau_j)$, $j = 1, \ldots, n-1$, born at time $\tau_j < \tau_n$ and located at x_j, independently with probability

$$\rho\left(\frac{\tau_n \, \mathrm{d}_1(x_n, x_j)^d}{\beta(\tau_n/\tau_j)^\gamma}\right), \tag{2}$$

where

$$\mathrm{d}_t(x, y) := \min\left\{|x - y + u| : u \in \{-\sqrt[d]{t}, 0, \sqrt[d]{t}\}^{\times d}\right\}$$

denotes the standard metric on the d-dimensional torus of volume t.

- Whenever an arc $\mathbf{x}_n \to \mathbf{x}_j$ is formed, the older vertex $\mathbf{x}_j$ forms an arc to $\mathbf{x}_n$ independently with probability $(\tau_j/\tau_n)^\Gamma$.

We think of $(\mathscr{D}_t : t \geq 0)$ as a growing network of social media users and of arcs representing a 'follow'-type relation. Users/accounts arrive over time and establish arcs to already existing accounts based on general popularity (modelled by age) and preexisting preference (modelled by spatial closeness). If an established account receives a new follower, they may decide if they wish to reciprocate the relation, which happens with probability 'birth time quotient to the Γ'. Since $1^\Gamma = 1$, if the two accounts are approximately of the same age, the occurrence of a reverse arc is quite probable. However, the older the old vertex is compared to the younger one, the less likely the presence of the reverse arc becomes. Hence, it is a rare event for a popular (typically old) account to follow one of their many younger followers, which captures the dynamics of real world social media networks.

Since the precise ordering of the birth times becomes irrelevant in the following, we go back to the previously used notation and denote vertices by $\mathbf{x} = (x, t_x)$, where $x \in [-1/2, 1/2)^d$ denotes the location and $t_x \in (0, \infty)$ the birth time of the vertex $\mathbf{x}$. To relate $(\mathscr{D}_t : t \geq 0)$ to $\mathscr{D}$ we need a two step procedure of *rescaling* and *localisation*.

Rescaling. Define for any given $t \in (0, \infty)$ the rescaling map

$$h_t: \quad \left[-\tfrac{1}{2}, \tfrac{1}{2}\right)^d \times (0, t) \to \left[-\tfrac{t^{1/d}}{2}, \tfrac{t^{1/d}}{2}\right)^d \times (0, 1)$$
$$(y, s) \mapsto \left(t^{1/d}y, \tfrac{s}{t}\right).$$

For fixed $t > 0$, denote the vertex set of $\mathscr{D}_t$ by $\mathcal{X}_t$. The rescaling map h_t acts on the point set $\mathcal{X}_t$ and is extended canonically to the respective geometric digraphs graphs by defining $h_t(\mathscr{D}_t)$ to be the graph with vertex set $h_t(\mathcal{X}_t)$ where an arc $h_t(\mathbf{x}) \to h_t(\mathbf{y})$ is present if and only if $\mathbf{x} \to \mathbf{y}$ is present in $\mathscr{D}_t$. Note that $h_t(\mathcal{X}_t)$ is distributed as a unit intensity Poisson point process on $[-\sqrt[d]{t}/2, \sqrt[d]{t}/2)^d \times (0, 1)$, i.e. its points are located on the d-dimensional torus of volume t and carry birth times in $(0, 1)$. Moreover, for each $t_y < t_x < t$

$$\rho\left(\frac{\frac{t_x}{t} \, d_t(t^{1/d}x, t^{1/d}y)^d}{\beta\left(\frac{t_x/t}{t_y/t}\right)^\gamma}\right) = \rho\left(\frac{t_x \, d_1(x,y)^d}{\beta(t_x/t_y)^\gamma}\right) \quad \text{as well as} \quad \left(\frac{t_y/t}{t_x/t}\right)^\Gamma = \left(\frac{t_y}{t_x}\right)^\Gamma,$$

hence $h_t(\mathscr{D}_t)$ agrees in law with a restriction of $\mathscr{D}$ to vertices located in the spatial domain $[-\sqrt[d]{t}/2, \sqrt[d]{t}/2)^d$ with periodic boundary conditions. Note that this equality in law only holds for fixed t and does not extend to the process level. For instance, every given vertex in $\mathscr{D}$ has a fixed finite indegree, but the indegree of a fixed vertex in $\mathscr{D}_t$ diverges, as more and more vertices arrive cf. [21].

Localisation. We are interested in the long time behaviour of the graphs $(\mathscr{D}_t : t \geq 0)$ as seen from a *typical* vertex. Almost surely, there will be at least one vertex in the system if t is sufficiently large, hence from now on we work conditionally on the event that $\mathscr{D}_t$ is not the empty graph. For each[2] sufficiently large t, we choose a root vertex $\mathbf{o}_t$ uniformly at random and perform a shift of spatial coordinates such that $\mathbf{o}_t$ is located at the origin $0 \in [1/2, 1/2)^d$. We denote the resulting rooted geometric digraph by $(\mathscr{D}_t, \mathbf{o}_t)$. Extending the map h_t to rooted digraphs in the obvious way, it is not difficult to see that $h_t(\mathscr{D}_t, \mathbf{o}_t)$ corresponds to a version of $h_t(\mathscr{D}_t)$ with an additional vertex located at the origin $0 \in [-\sqrt[d]{t}/2, \sqrt[d]{t}/2)^d$.

The weak convergence theory of point processes, see e.g. [15], now strongly suggests that the rescaled and localised family of random geometric digraphs $(h_t(\mathscr{D}_t, \mathbf{o}_t) : t > 0)$ converges in distribution to a variant $(\mathscr{D}, \mathbf{o})$ of $\mathscr{D}$ with an additional vertex $\mathbf{o} = (0, t_0)$ at the origin, cf. [28] for related general results for finite domains with non-periodic boundary conditions in the undirected setup. A formal proof of this result can be given along the same lines as [29, Proposition 5]. As a corollary, we obtain a distributional limit theorem in D_*, the space of

---

[2] Clearly, it suffices to update the root vertex only at the birth times of new vertices, since we are merely interested in the one dimensional marginal distributions of the resulting family of rooted graphs.

rooted simple digraphs[3] with metric

$$d_*\big((G_1,o_1),(G_2,o_2))\big) = \frac{1}{1+\sup\{r : (G_1,o_1) \overset{r}{\simeq} (G_2,o_2)\}},$$

where we write $(G_1,o_1) \overset{r}{\simeq} (G_2,o_2)$ if there exists a digraph isomorphism mapping the r-neighbourhood of the root o_1 in G_1 to the r-neighbourhood of the root o_2 in G_2.

Theorem 1 (Local limit). *For $t \in \mathbb{N}$, let $(\mathcal{D}_t, \mathbf{o}_t) = h_t(\mathscr{D}_t, \mathbf{o}_t) \in D_*$ denote the rooted simple digraphs obtained from the rooted geometric graphs $(\mathscr{D}_t, \mathbf{o}_t)$ via rescaling. Let further denote $(\mathscr{D}, \mathbf{o})$ the rooted digraph obtained from generating $\mathscr{D}$ with an additional (root) vertex located at the origin. Then*

$$(\mathcal{D}_t, \mathbf{o}_t) \xrightarrow[t\to\infty]{} (\mathscr{D}, \mathbf{o}) \text{ in distribution.}$$

In the following section, we calculate local metrics for $\mathscr{D}$. By virtue of Theorem 1, the corresponding metrics for $(\mathscr{D}_t : t \geq 0)$ converge to the corresponding limit values for (the rooted version of) $\mathscr{D}$.

3   Local Properties

In this section, we establish important local properties of $(\mathscr{D}, \mathbf{o})$. Here, $\mathbf{o}$ denotes the additionally added root vertex which can be seen as a typical vertex in $\mathscr{D}$ as explained in the previous paragraph. We call a property local if it only depends on a bounded graph neighbourhood of $\mathbf{o}$. Formally, such properties have representations via continuous functionals on the space D_*. We begin by identifying the degree distribution of $\mathbf{o}$. Afterwards, we discuss clustering metrics.

In the following, we use the established notation $f \asymp g$ for non-negative functions to indicated that $f(x)/g(x)$ is bounded from zero and infinity.

3.1   Degree Distribution

For a given vertex $\mathbf{x}$ we denote by

$$\mathscr{N}^{\text{in}}(\mathbf{x}) := \{\mathbf{y} \in \mathcal{X} : \mathbf{y} \to \mathbf{x}\}$$

the vertices sending arcs to $\mathbf{x}$ in $\mathscr{D}$ and by $\sharp\mathscr{N}^{\text{in}}(\mathbf{x})$ its indegree. If $\mathbf{x} = \mathbf{o}$, we simply write $\mathscr{N}^{\text{in}}$. For outgoing arcs and outdegree, we use the analogous notations $\mathscr{N}^{\text{out}}(\mathbf{x})$, resp. $\mathscr{N}^{\text{out}}$, and $\sharp\mathscr{N}^{\text{out}}(\mathbf{x})$, resp. $\sharp\mathscr{N}^{\text{out}}$.

---

[3] In fact, D_* is the quotient space of equivalence classes of rooted graphs up to graph isomorphism, but we do not distinguish between a graph and its isomorphism class here. For more background on D_* and local weak convergence see [1].

Lemma 1 (Degree distribution).

(a) For the indegree of **o** *in* $\mathscr{D} = \mathscr{D}[\beta, \gamma, \delta, \Gamma]$, *we have for all* $\gamma \in (0, 1)$ *and* $\Gamma > 0$ *that*

$$\mathbb{P}_0(\sharp \mathscr{N}^{\mathrm{in}} = k) = k^{-1-1/\gamma + o(1)}, \quad \text{as } k \uparrow \infty, \text{ and}$$

(b) for the outdegree of **o** *in* $\mathscr{D} = \mathscr{D}[\beta, \gamma, \delta, \Gamma]$, *we have for all* $\gamma \in (0, 1)$ *that*
  (i) if $\Gamma > \gamma$, *then* $\sharp \mathscr{N}^{\mathrm{out}}$ *is Poisson distributed with parameter* $\beta/(\Gamma-\gamma)$,
  (ii) if $0 < \Gamma < \gamma$, *then* $\sharp \mathscr{N}^{\mathrm{out}}$ *is mixed Poisson distributed with mixing density*

$$f^{\mathrm{out}}(\lambda) = (\gamma - \Gamma)^{-(2+1/(\gamma-\Gamma))} \left(\tfrac{\beta}{\lambda}\right)^{1+1/(\gamma-\Gamma)} \mathbb{1}_{(\beta/(\gamma-\Gamma),\infty)}(\lambda)$$

and therefore

$$\mathbb{P}_0(\sharp \mathscr{N}^{\mathrm{out}} = k) = k^{-1-1/(\gamma-\Gamma)+o(1)}, \quad \text{as } k \uparrow \infty.$$

Proof. The incoming edges of **o** are all edges to younger neighbours of **o** in $\mathscr{G}$ plus the edges to older neighbours where a reciprocal arc has been added. From [21, Proposition 4.1(d)], the number of younger neighbours in $\mathscr{G}$ from $(0, u)$ (i.e. the root's mark is given by $U_0 = u$) is mixed Poisson with mixing density

$$f^{\mathrm{in}}(\lambda) \asymp \lambda^{-1-1/\gamma}.$$

The older incoming neighbours of $(0, u)$ form a Poisson process on $\mathbb{R}^d \times (0, u)$ with intensity

$$\left(\tfrac{s}{u}\right)^{\Gamma} \rho(\beta^{-1} s^{\gamma} u^{1-\gamma} |x|^d) \mathrm{d}x\mathrm{d}s.$$

Since $(s/u)^{\Gamma} \leq 1$, the number of such neighbours is at most Poisson distributed with parameter $\beta/(1-\gamma)$ [21, Proposition 4.1(c)]. Hence, the indegree of **o** in $\mathscr{D}$ is bounded from below by the number of younger neighbours of **o** in $\mathscr{G}$ and from above by the number of younger neighbours of **o** in $\mathscr{G}$ plus an independent Poisson distributed random variable. As the number of younger neighbours of **o** in $\mathscr{G}$ is heavy tailed with power-law exponent $\tau = 1 + 1/\gamma$, cf. [21, Lemma 4.4] both bounds are of the same order, proving (a).

Similarly, the outgoing neighbours of **o** in $\mathscr{D}$ are the older neighbours of **o** in $\mathscr{G}$ plus the younger ones where a reciprocal arc has been added. The number of the first type is again Poisson distributed independently of the root's mark. For fixed mark $U_0 = u$, the latter form a Poisson process on $\mathbb{R}^d \times (u, 1)$ with intensity

$$\left(\tfrac{u}{s}\right)^{\Gamma} \rho\big(\beta^{-1} u^{\gamma} s^{1-\gamma} |x|^d\big) \, \mathrm{d}x \, \mathrm{d}s.$$

The expected number of such vertices is

$$\int_u^1 \mathrm{d}s \, \left(\tfrac{u}{s}\right)^{\Gamma} \int_{\mathbb{R}^d} \mathrm{d}x \, \rho(\beta^{-1} u^{\gamma} s^{1-\gamma} |x|^d)$$

$$\asymp \beta u^{-\gamma} \int_u^1 \left(\tfrac{u}{s}\right)^{\Gamma} s^{\gamma-1} \mathrm{d}s \asymp \beta u^{\Gamma-\gamma} \int_u^1 s^{\gamma-\Gamma-1} \, \mathrm{d}s \asymp \frac{\beta}{|\gamma - \Gamma|}(1 \vee u^{-\gamma+\Gamma}).$$

Hence, if $\Gamma > \gamma$, the outdegree is Poisson distributed with a parameter independent of u. If $\Gamma < \gamma$, then the outdegree is mixed Poisson distributed and

$$\mathbb{P}_0(\sharp\mathcal{N}^{\mathrm{out}} = k) = \int_0^1 e^{\frac{\beta u^{\Gamma-\gamma}}{\gamma-\Gamma}} \frac{\left(\frac{\beta u^{\Gamma-\gamma}}{\gamma-\Gamma}\right)^k}{k!} du \asymp \int_{\beta/(\gamma-\Gamma)}^\infty e^{-\lambda}\frac{\lambda^k}{k!} f^{\mathrm{out}}(\lambda)d\lambda.$$

Hence, by Stirling's formula, we have

$$\mathbb{P}_0(\sharp\mathcal{N}^{\mathrm{out}} = k) \asymp \frac{1}{\Gamma(k+1)}\int_0^\infty e^{-\lambda}\lambda^{k-1/(\gamma-\Gamma)-1}d\lambda \asymp k^{-1-1/(\gamma-\Gamma)},$$

as $k \to \infty$, concluding the proof.

3.2 Clustering

In this section, we discuss two clustering measures. Firstly, the *local friend clustering coefficient* and secondly the *interest clustering number*. The idea of the friend clustering coefficient is that two friends of a typical vertex are more likely to be friends with each other than two general typical vertices. Here, a 'friendship' denotes a reciprocal connection, and the corresponding clustering coefficient measures the density of strongly connected triangles. This coincides with our motivation of the model that the network consists mainly of typical (young) vertices and some very influential (old) vertices. While the influential vertices are those with much larger in than outdegree, the typical vertices should tend to form triangles, just like in the age-dependent random connection model in the undirected case. The friend clustering coefficient is a straightforward adaptation from the undirected setting, cf. [21].

The idea of interest clustering was proposed in Trolliet et al. [35] explicitly for directed social networks. Our localised adaptation of their coefficient appears here for the first time. It is based on the idea that in an (online) social network, the clustering is also driven by common interests. Whereas the friend clustering coefficient is a metric intended to capture purely the social aspect of network formation, the interest clustering number combines the social with the informational aspect.

We first define both clustering metrics only in the infinite limit model $(\mathcal{D}, \mathbf{o})$ and provide integral representations for them and then consider the finite models $(\mathcal{D}_t : t \geq 0)$.

The Friend Clustering Coefficient. In this section, we call two given vertices $\mathbf{x}$ and $\mathbf{y}$ *friends* in $\mathcal{D}$ if $\mathbf{x} \leftrightarrow \mathbf{y}$. Let $\overleftrightarrow{\mathcal{V}}_2$ the set of all vertices having at least two friends in $\mathcal{D}$. For $\mathbf{x} \in \overleftrightarrow{\mathcal{V}}_2$, we define the *friend clustering coefficient* as

$$c^{\mathrm{fc}}(\mathbf{x}) := \frac{\sum_{\mathbf{y},\mathbf{z}\in\mathcal{X}:t_y>t_z} \mathbb{1}_{\{\mathbf{x}\leftrightarrow\mathbf{y}\}}\mathbb{1}_{\{\mathbf{x}\leftrightarrow\mathbf{z}\}}\mathbb{1}_{\{\mathbf{y}\leftrightarrow\mathbf{z}\}}}{\binom{\sharp(\mathcal{N}^{\mathrm{out}}(\mathbf{x})\cap\mathcal{N}^{\mathrm{in}}(\mathbf{x}))}{2}}.$$

If $\mathbf{x} \notin \overleftrightarrow{\mathcal{V}}_2$, we set its friend clustering coefficient to be zero.

Lemma 2 (Friend clustering). *For all $\beta > 0, \gamma \in (0,1), \delta > 1$, and $\Gamma \geq 0$, we have*

$$\mathbb{E}_0 c^{fc}(\mathbf{o}) = \int_0^1 du\, \mathbb{P}(\mathbf{Y}^{(u)} \leftrightarrow \mathbf{X}^{(u)}) \mathbb{P}_{(0,u)}\Big(\bigcup_{k \geq 2} F_{(0,u)}(k) \Big) > 0,$$

where $F_{(o,u(k))}$ is the event that the root (o,u) has k friends and $\mathbf{X}^{(u)}$ and $\mathbf{Y}^{(u)}$ are two independent random variables distributed according to the normalised measure $\lambda_u^f / \lambda_u^f(\mathbb{R}^d)$ with

$$\lambda_u^f = \Big(\big(\tfrac{s}{u}\big)^{\Gamma} \rho(\beta^{-1} s^{\gamma} u^{1-\gamma} |x|^d) \mathbb{1}_{\{s<u\}} + \big(\tfrac{u}{s}\big)^{\Gamma} \rho(\beta^{-1} s^{1-\gamma} u^{\gamma} |x|^d) \mathbb{1}_{\{s \geq u\}} \Big)\, ds\, dx.$$

We do not give the proof here as it works analogously to the undirected ARCM [21, Theorem 5.1].

The Interest Clustering Number. Consider two vertices $\mathbf{x}, \mathbf{y} \in \overrightarrow{\mathscr{V}}_2$ and define the quantity

$$c^{\mathrm{ic}}(\mathbf{y}|\mathbf{x}) = \begin{cases} 0, & \text{if } \sharp(\overrightarrow{N}(\mathbf{y}) \cap \overrightarrow{N}(\mathbf{x})) < 2, \\ \dfrac{\sum_{u,v \in \mathcal{X}} \mathbb{1}_{\{y \to u, y \to v\}} \mathbb{1}_{\{x \to u, x \to v\}}}{\sum_{u,v \in \mathcal{X}} \mathbb{1}_{\{y \to u \text{ or } y \to v\}} \mathbb{1}_{\{x \to u, x \to v\}}}, & \text{otherwise.} \end{cases}$$

Note that $c^{\mathrm{ic}}(\mathbf{y}|\mathbf{x})$ is at most 1. Let $\mathbb{P}_{0,x}$ denote the law of the digraph $\mathscr{D}$ constructed from a vertex set with two additional vertices $\mathbf{o}, \mathbf{x}$ located at 0 and $x \in \mathbb{R}^d$, respectively. Then we call

$$n^{\mathrm{ic}}(\mathbf{o}) = \int_{\mathbb{R}^d} \mathbb{E}_{0,x} c^{\mathrm{ic}}(\mathbf{x}|\mathbf{o})\, dx$$

the *interest clustering number* of $\mathscr{D}$. The number $n^{\mathrm{ic}}(\mathbf{o})$ can be interpreted as a localised version of the 'interest clustering coefficient' proposed in [35] for directed graphs derived from social and information network data. An important difference is that $n^{\mathrm{ic}}(\mathbf{o}) \in (0, \infty)$ is not a normalised quantity. A large value of $n^{\mathrm{ic}}(\mathbf{o})$ implies that typical vertices who share a common interest (i.e. both send an arc to a third vertex) are likely to have further common interests, whereas a small value of $n^{\mathrm{ic}}(\mathbf{o})$ indicates that interests are formed more or less independently of each other. To formulate our result regarding interest clustering in $\mathscr{D}$, we need the two numbers $\mu_{11}(x)$ and $\mu_{10}(x)$ given by

$$\mu_{11}(x) = \int_{\mathbb{R}^d} \int_0^1 \int_0^1 \int_0^{u \wedge s} \big(\tfrac{t}{u}\big)^{\Gamma} \rho(\beta^{-1} t^{\gamma} u^{1-\gamma} |y|^d) \big(\tfrac{t}{s}\big)^{\Gamma} \rho(\beta^{-1} t^{\gamma} s^{1-\gamma} |x-y|^d)\, dt$$

$$+ \mathbb{1}_{\{u<s\}} \int_u^s \rho(\beta^{-1} u^{\gamma} t^{1-\gamma} |y|^d) \big(\tfrac{t}{s}\big)^{\Gamma} \rho(\beta^{-1} t^{\gamma} s^{1-\gamma} |x-y|^d)\, dt$$

$$+ \mathbb{1}_{\{s<u\}} \int_s^u \big(\tfrac{t}{u}\big)^{\Gamma} \rho(\beta^{-1} t^{\gamma} u^{1-\gamma} |y|^d) \rho(\beta^{-1} s^{\gamma} t^{1-\gamma} |x-y|^d)\, dt$$

$$+ \int_{s \vee u}^1 \rho(\beta^{-1} u^{\gamma} t^{1-\gamma} |y|^d) \rho(\beta^{-1} s^{\gamma} t^{1-\gamma} |x-y|^d)\, dt\, du\, ds\, dy,$$

and

$$
\mu_{l0}(x) = \int_{\mathbb{R}^d} \int_0^1 \int_0^1 \int_0^{u \wedge s} \left(\tfrac{t}{u}\right)^\Gamma \rho\big(\beta^{-1} t^\gamma u^{1-\gamma} |y|^d\big) \left(\tfrac{t}{s}\right)^\Gamma \rho\big(\beta^{-1} t^\gamma s^{1-\gamma} |x-y|^d\big) \, dt
$$

$$
+ \mathbb{1}_{\{u<s\}} \int_u^s \rho\big(\beta^{-1} u^\gamma t^{1-\gamma} |y|^d\big) \left(1 - \left(\tfrac{t}{s}\right)^\Gamma \rho\big(\beta^{-1} t^\gamma s^{1-\gamma} |x-y|^d\big)\right) dt
$$

$$
+ \mathbb{1}_{\{s<u\}} \int_s^u \left(\tfrac{t}{u}\right)^\Gamma \rho\big(\beta^{-1} t^\gamma u^{1-\gamma} |y|^d\big) \left(1 - \rho\big(\beta^{-1} s^\gamma t^{1-\gamma} |x-y|^d\big)\right) dt
$$

$$
+ \int_{s \vee u}^1 \rho\big(\beta^{-1} u^\gamma t^{1-\gamma} |y|^d\big) \left(1 - \rho\big(\beta^{-1} s^\gamma t^{1-\gamma} |x-y|^d\big)\right) dt \, du \, ds \, dy.
$$

Lemma 3 (Interest clustering). *For all $\beta > 0, \gamma \in (0,1), \delta > 1$, and $\Gamma \geq 0$, we have*

$$
n^{ic}(\mathbf{o}) = \int_{\mathbb{R}^d} dx \sum_{k=2}^\infty e^{-\mu_{ll}(x)} \frac{\mu_{ll}(x)^k}{k!}
$$

$$
- 2\mu_{l0}(x) \int_0^\infty ds \, e^{\mu_{l0}(x)(e^{-2s}-1)} \left(e^{\mu_{ll}(x)(e^{-s}-1)-s} - e^{-\mu_{ll}(x)-s} - e^{-\mu_{ll}(x)-2s} \mu_{ll}(x) \right).
$$

Proof. Under $\mathbb{P}_{0,x}$, the number Y_{ll} of vertices connected to both $\mathbf{o}$ and $\mathbf{x}$ is Poisson distributed with parameter $\mu_{ll} = \mu_{ll}(x)$ and the number Y_{l0} of vertices connected to $\mathbf{o}$ but not to $\mathbf{x}$ is Poisson distributed with parameter $\mu_{l0} = \mu_{l0}(x)$ and independent of Y_{ll}. On the event $\{Y_{ll} > 1\}$, we have

$$
\frac{\sum_{\mathbf{u},\mathbf{v} \in \mathcal{X}} \mathbb{1}_{\{y \to u, y \to v\}} \mathbb{1}_{\{x \to u, x \to v\}}}{\sum_{\mathbf{u},\mathbf{v} \in \mathcal{X}} \mathbb{1}_{\{y \to u \text{ or } y \to v\}} \mathbb{1}_{\{x \to u, x \to v\}}} = \frac{Y_{ll}(Y_{ll}-1)}{Y_{ll}(Y_{ll}-1) + 2Y_{ll}Y_{l0}} = \frac{Y_{ll}-1}{Y_{ll}-1+2Y_{l0}}.
$$

Hence,

$$
\mathbb{E}_{0,x} c^{ic}(\mathbf{x}|\mathbf{o}) = \sum_{k=2}^\infty e^{-\mu_{ll}} \frac{\mu_{ll}^k}{k!} (k-1) \mathbb{E}_{0,x} \left[\frac{1}{k-1+2Y_{l0}} \right]
$$

$$
= \sum_{k=2}^\infty e^{-\mu_{ll}} \frac{\mu_{ll}^k}{k!} (k-1) \int_0^\infty \mathbb{E}_{0,x} e^{-s(k-1+2Y_{l0})} \, ds
$$

$$
= \sum_{k=2}^\infty e^{-\mu_{ll}} \frac{\mu_{ll}^k}{k!} (k-1) \int_0^\infty e^{\mu_{l0}(e^{-2,s}-1)-s(k-1)} \, ds
$$

$$
= \sum_{k=2}^\infty e^{-\mu_{ll}} \frac{\mu_{ll}^k}{k!} \left(1 + \int_0^\infty e^{-s(k-1)} \frac{d}{ds} e^{\mu_{l0}(e^{-2s}-1)} \, ds \right).
$$

Since

$$
\frac{d}{ds} e^{\mu_{l0}(e^{-2s}-1)} = -2\mu_{l0} e^{\mu_{l0}(e^{-2s}-1)-2s},
$$

we have

$$
\mathbb{E}_{0,x} c^{\mathrm{ic}}(\mathbf{x}|\mathbf{o}) = \sum_{k=2}^{\infty} \mathrm{e}^{-\mu_{\|}} \frac{\mu_{\|}^{k}}{k!} \left(1 - 2\mu_{\mathsf{IO}} \int_{0}^{\infty} \mathrm{e}^{\mu_{\mathsf{IO}}(\mathrm{e}^{-2s}-1)-s(k+1)} \, \mathrm{d}s \right)
$$

$$
= \sum_{k=2}^{\infty} \mathrm{e}^{-\mu_{\|}} \frac{\mu_{\|}^{k}}{k!} - 2\mu_{\mathsf{IO}} \int_{0}^{\infty} \mathrm{e}^{\mu_{\mathsf{IO}}(\mathrm{e}^{-2s}-1)} \sum_{k=2}^{\infty} \mathrm{e}^{-\mu_{\|}} \frac{\mu_{\|}^{k}}{k!} \mathrm{e}^{-s(k+1)} \, \mathrm{d}s.
$$

The last sum can be rewritten as

$$
\sum_{k=2}^{\infty} \mathrm{e}^{-\mu_{\|}} \frac{\mu_{\|}^{k}}{k!} \mathrm{e}^{-s(k+1)} = \mathrm{e}^{\mu_{\|}(\mathrm{e}^{-s}-1)-s} - \mathrm{e}^{-\mu_{\|}-s} - \mathrm{e}^{-\mu_{\|}-2s} \mu_{\|},
$$

and integrating the whole expression for $\mathbb{E}_{0,x} c^{\mathrm{ic}}(\mathbf{x}|\mathbf{o})$ over $x \in \mathbb{R}^d$ now yields the representation given in the lemma.

Clustering in $(\mathscr{D}_t : t \geq 0)$. It is straightforward to see, that both metrics are positive and *local*, i.e. continuous with respect to d_*. Hence Theorem 1 implies that the finite systems $(\mathscr{D}_t)$ asymptotically display both forms of clustering, in the sense that the corresponding statistically averaged metrics converge to the metrics of the limit system $\mathscr{D}$. To arrive at the same conclusion by defining the average clustering metrics ad hoc for the finite model and evaluate suitable limiting expressions analytically would require a more careful mathematical treatment. The use of the general local convergence framework allows us to avoid these complications.

Theorem 2 (Asymptotics of average clustering metrics). *We have that*

$$
\frac{1}{\sharp V(\mathscr{D}_t)} \sum_{\mathbf{x} \in V(\mathscr{D}_t)} c^{fc}(\mathbf{x}) \to \mathbb{E}_0 c^{fc}(\mathbf{o}) \ \text{in probability,}
$$

and that

$$
\frac{1}{\sharp V(\mathscr{D}_t)} \sum_{\mathbf{x} \in V(\mathscr{D}_t)} \sum_{\mathbf{y} \in V(\mathscr{D}_t)} c^{ic}(\mathbf{y}|\mathbf{x}) \to n^{ic}(\mathbf{o}) \ \text{in probability.}
$$

Proof (Sketch). The first statement is immediate from Theorem 1, upon noticing that the averaging corresponds to choosing a uniform root at which to evaluate $c^{\mathrm{fc}}(\cdot)$ and taking expectations. The same holds true for the outer averaging in the second expression; an application of Campbell's formula [31] then reduces the inner summation to an integral as in the definition of $n^{\mathrm{ic}}(\mathbf{o})$.

4   Directed Percolation

Originally introduced by Broadbent and Hammersley in 1957 [9], percolation has drawn a lot of attention from the mathematical community and is widely studied until today. The most fundamental question in percolation theory is whether an infinite connected component (*cluster*) exists. If such a component exists, the graph can be seen as well connected whereas if no such component exists the graph decomposes into a collection of disjoint finite clusters. Since the existence of an infinite cluster is monotone in the edge density β and a 0-1-event, one needs to only establish the existence of a critical intensity $\beta_c \in (0, \infty)$ such that an infinite component is almost surely present in the graph if $\beta > \beta_c$ but almost surely absent if $\beta < \beta_c$. For undirected translation invariant models in $\mathbb{R}^d$ it is well established that there is at most one unique infinite component present [11, 30]. If $\beta_c = 0$, the graph is also referred to being *robust*. Percolation was studied for various models since its introduction, see e.g. [16, 18, 22–24, 32]. A related but typically more difficult question is whether or not a growing sequence of graph contains a (unique) component of linear size. Whether in general percolation in the local limit has a bearing on the existence of linear components in the approximating graph sequence has recently been investigated in [27].

In our directed setting, there are now two types of connected components, weak and strong ones. Let us denote by $\mathbf{x} \longrightarrow \mathbf{y}$ the event that there exists a directed path from $\mathbf{x}$ to $\mathbf{y}$ in $\mathscr{D}$. Then, $\mathbf{x}$ and $\mathbf{y}$ belong to the same *weakly connected component* if either $\mathbf{x} \longrightarrow \mathbf{y}$ or $\mathbf{y} \longrightarrow \mathbf{x}$. On the contrary, $\mathbf{x}$ and $\mathbf{y}$ belong to a *strongly connected component* if $\mathbf{x} \longrightarrow \mathbf{y}$ and $\mathbf{y} \longrightarrow \mathbf{x}$ which we denote from now on as $\mathbf{x} \longleftrightarrow \mathbf{y}$. This gives rise to three components of the root (resp. a given vertex) to consider:

$$\vec{\mathcal{C}} = \{\mathbf{x} \in \mathcal{X} : \mathbf{o} \longrightarrow \mathbf{x}\}, \ \overleftarrow{\mathcal{C}} = \{\mathbf{x} \in \mathcal{X} : \mathbf{x} \longrightarrow \mathbf{o}\}, \text{ and } \overleftrightarrow{\mathcal{C}} = \{\mathbf{x} \in \mathcal{X} : \mathbf{o} \longleftrightarrow \mathbf{x}\}.$$

In this article, we focus on the weak-connectedness event

$$\{\sharp \vec{\mathcal{C}} = \infty\} := \{\mathbf{o} \longrightarrow \infty\}$$

only, other percolation questions are left for future work. The event $\mathbf{o} \longrightarrow \infty$ can be interpreted as the situation that news spread through the networks $(\mathscr{D}_t)$ by the most influential vertices can reach a positive proportion of the network. In order to make this rigorous however, one would need a convergence result for a weak giant component similar to the undirected case. Since this is not a local event, Theorem 1 cannot be applied and an extension of [27] for the directed case is needed. Note however, that in digraphs the existence of a large weakly connected component can occur even if the local limit does not weakly percolate, see e.g. [33], which cannot happen in undirected graphs, hence the situation is more complex as in the undirected setting.

Similar to the undirected case, we are interested in the critical intensity

$$\vec{\beta}_c := \vec{\beta}_c(\gamma, \delta, \Gamma) = \sup\{\beta > 0 : \mathbb{P}_0\{\mathbf{o} \longrightarrow \infty\} = 0\}.$$

If one restricts the graph to edges that point in both directions and vertices with birth time larger than $1/2$, it is easy to see that this graph can be compared with a undirected (long-range) random connection model for which the existence of an infinite component in dimensions $d \geq 2$ and for $\delta \in (1, 2)$ in $d = 1$ is well-known [32]. Hence, we immediately infer $\vec{\beta}_c < \infty$ in these cases. Therefore, we deal with the question of positivity of the critical intensity here.

Theorem 3 (Positivity of critical intensity). *Consider the DARCM $\mathscr{D} = \mathscr{D}[\beta, \gamma, \delta, \Gamma]$ with $\beta > 0$, $\gamma \in (0, 1)$, $\delta > 1$, and $\Gamma \geq 0$.*

(i) If $\gamma < (\delta+\Gamma)/(\delta+1)$, then $\vec{\beta}_c > 0$ and

(ii) if $\gamma > (\delta+\Gamma)/(\delta+1)$, then $\vec{\beta}_c = 0$.

The proof further elaborates the ideas for the undirected case in [22]. There, it was shown that the most promising strategy for building a long path is to use young intermediate 'connectors' to connect old vertices and if $\gamma > \delta/(\delta+1)$ the age's influence is strong enough compared to the geometric restrictions such that this strategy can be repeated indefinitely with positive probability regardless of the edge intensity. We will adapt the strategy of [22] to the directed setting. Let us write

$$ \mathbf{x} \xrightarrow[\mathbf{x,y}]{2} \mathbf{y} $$

for the event that $\mathbf{x}$ is connected to $\mathbf{y}$ by a directed path of length 2 where the intermediate vertex is younger than both $\mathbf{x}$ and $\mathbf{y}$. Key to the proof of Theorem 3 is the following lemma.

Lemma 4 (Two-connection-lemma). *Consider $\mathscr{D} = \mathscr{D}[\beta, \gamma, \delta, \Gamma]$ with $\beta > 0$, $\gamma \in (0, 1)$, $\delta > 1$, and $\Gamma \geq 0$.*

(a) Assume $\gamma < (\delta+\Gamma)/(\delta+1)$. Let $\mathbf{x} = (x, t)$ and $\mathbf{y} = (y, s)$ be two given vertices that satisfy $|x - y|^d \geq \beta(t \wedge s)^{-\gamma}(s \vee t)^{\gamma-1}$. Then we have

$$ \mathbb{P}_{\mathbf{x,y}}\left(\mathbf{x} \xrightarrow[\mathbf{x,y}]{2} \mathbf{y} \right) \leq \mathbb{E}_{\mathbf{x,y}}\left[\sharp\{\mathbf{z} = (z, u) : u > t \text{ and } \mathbf{x} \to \mathbf{z} \to \mathbf{y}\} \right] $$
$$ \leq \beta C \, \mathbb{P}_{\mathbf{x,y}}(\mathbf{x} \to \mathbf{y}). \tag{3} $$

for some $C > 0$ only depending on γ, δ, Γ, and the dimension d.

(b) Let $\gamma > (\delta+\Gamma)/(\delta+1)$ and fix two interdependent constants

$$ \alpha_1 \in \left(1, \tfrac{\gamma-\Gamma}{\delta(1-\gamma)}\right) \text{ and } \alpha_2 \in \left(\alpha_1, \tfrac{\alpha_1(\gamma\delta-1)+\gamma-\Gamma}{\delta-1}\right). $$

Let $\mathbf{x} = (x, t)$ be a given vertex with $t < 1/2$ and define the event

$$ E(\mathbf{x}) = \left\{ \exists \mathbf{y} = (y, s) : s < t^{\alpha_1}, |x - y|^d < t^{-\alpha_2} \text{ and } \mathbf{x} \xrightarrow[\mathbf{x,y}]{2} \mathbf{y} \right\}. $$

For each $\beta > 0$ there exists some $a < 0$ such that $\mathbb{P}_{\mathbf{x}}(E(\mathbf{x})) \geq 1 - e^{-t^a}$.

Proof. We start by proving (a) and focus on the case $s < t$. The other case works analogously. Observe that the first inequality in (3) is simply a moment bound, and we hence focus on the second inequality. By our assumptions on the distance of $\mathbf{x}$ and $\mathbf{y}$ we have

$$\mathbb{E}_{\mathbf{x},\mathbf{y}}\big[\sharp\{\mathbf{z} = (z,u) : u > t \text{ and } \mathbf{x} \to \mathbf{z} \to \mathbf{y}\}\big]$$

$$\leq \int_{\mathbb{R}^d} dz \int_t^1 du \left(\tfrac{t}{u}\right)^{\Gamma} \rho\big(\tfrac{1}{\beta} t^{\gamma} u^{1-\gamma} |x - z|^d\big) \rho\big(\tfrac{1}{2^d\beta} s^{\gamma} u^{1-\gamma} |x - y|^d\big) \mathbb{1}_{\{|y-z| > \frac{|x-y|}{2}\}}$$

$$+ \int_{\mathbb{R}^d} dz \int_t^1 du \left(\tfrac{t}{u}\right)^{\Gamma} \rho\big(\tfrac{1}{2^d\beta} t^{\gamma} u^{1-\gamma} |x - y|^d\big) \rho\big(\tfrac{1}{\beta} s^{\gamma} u^{1-\gamma} |y - z|^d\big) \mathbb{1}_{\{|x-z| \geq \frac{|x-y|}{2}\}}$$

$$\leq \beta C \big(\tfrac{1}{\beta} s^{\gamma} t^{1-\gamma} |x - y|^d\big)^{-\delta} = \beta C \, \rho\big(\tfrac{1}{\beta} s^{\gamma} t^{1-\gamma} |x - y|^d\big),$$

where we integrated both integrals separately and have used that indicator functions are bounded by 1, a change of variables together with the integrability of ρ, as well as $\gamma < {(\delta+\Gamma)}/{(\delta+1)}$ and our distance assumption.

We also start the proof of (b) by calculating the expected number of vertices $\mathbf{y}$ being part of the event $E(\mathbf{x})$. Using similar arguments as above, it is straight forward to deduce that this expectation is lower bounded by

$$ct^{\Gamma - \gamma + \alpha_2(\delta-1) - \alpha_1(\gamma\delta-1)},$$

for some constant $c > 0$. The proof finishes with the observation that due to our choices of α_1 and α_2, we have $a := \Gamma - \gamma + \alpha_2(\delta - 1) - \alpha_1(\gamma\delta - 1) < 0$ and therefore $\mathbb{P}_{\mathbf{x}}(E(\mathbf{x})) \geq 1 - e^{-ct^a}$.

Proof (of Theorem 3). The proof works similarly to the proof carried out in [22] and shall hence only be sketched. Observe first, that the origin starts a path $\mathbf{o} \longrightarrow \infty$ using young connectors to connect to older and older vertices with strictly positive probability if the root itself is old enough by Lemma 4 (b), proving Part (ii).

To prove Part (i), we want to bound the probability that the root $\mathbf{o}$ starts a directed, short-cut free path of length n by a term that goes to 0 as $n \to \infty$ for small enough β. Here, a directed path $P = (\mathbf{x}_0, \dots, \mathbf{x}_n)$ is called *shortcut-free*, if $\mathscr{N}^{\text{in}}(\mathbf{x}_j) \cap P = \mathbf{x}_{j-1}$ and $\mathscr{N}^{\text{out}}(\mathbf{x}_j) \cap P = \mathbf{x}_{j+1}$ for all $j = 1, \dots, n-1$. Note that there always exists a directed shortcut-free path to infinity if there exists a directed path to infinity at all. To make use of the previous lemma, we work from now on in the graph $\widehat{\mathscr{D}}$ defined by taking $\mathscr{D}$ and adding each bi-directed arc $\mathbf{x} \leftrightarrow \mathbf{y}$ (if not already there) whenever $\mathbf{x}$ and $\mathbf{y}$ do not fulfil the distance condition of Lemma 4 (a). By definition, in a short-cut free path in $\widehat{\mathscr{D}}$ we always have

$$|x_i - x_j|^d > \beta(t_i \wedge t_j)^{-\gamma}(t_i \vee t_j)^{\gamma-1}, \qquad \text{whenever} \qquad |i - j| \geq 2.$$

From Lemma 4 (a), we infer that for all vertices fulfilling this distance condition, it is more probable to be connected by a direct arc than through a single connector. To make use of this fact, let us introduce the notion of a path's *skeleton*.

For a path P we call the collection of vertices with running minimum age from both sides the *skeleton* of P. That is, we start from the initial vertex (x_0, t_0)

and search for the first vertex (x_{j_1}, t_{j_1}) that has birth time $t_{j_1} < t_0$. Starting from this vertex again we search for the next vertex with smaller birth time still until we reach the oldest vertex of the path. Afterwards we do the same but starting from the last vertex of the path (x_n, t_n) and going backwards across the indices. Another possibility to identify the path's skeleton is the following: We call a vertex $\mathbf{x}_j \in P \setminus \{\mathbf{x}_0, \mathbf{x}_n\}$ local maximum if $t_j > t_{j-1}$ and $t_j > t_{j+1}$. Put differently, $\mathbf{x}_j$ is younger than its preceding and subsequent vertex. We now successively remove all local maxima from P as follows: First, take the local maximum in P with the greatest birth time, remove it from P and connect its former neighbours by a directed edge oriented from preceding to subsequent vertex. In the resulting path, we take the local maximum of greatest birth time and remove it, repeating until there is no local maximum left, see Fig. 1.

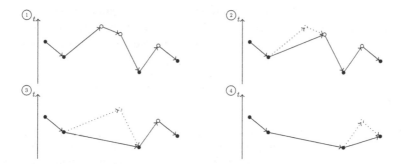

Fig. 1. Methodical sketch of the outlined step by step removal of local maxima to observe the skeleton path (the black vertices).

The idea is now the following. Between each two skeleton vertices, we remove all local maxima step by step and replace them by direct arcs. The probabilistic costs for each such replacement is the probability of an arc times βC. Since all combinatorics involved are also of exponential order in the subpath's length, cf. [22, Lemma 2.3], we infer for two given vertices $\mathbf{x}$ and $\mathbf{y}$ satisfying the distance condition

$$\mathbb{P}_{\mathbf{x},\mathbf{y}}\left(\mathbf{x} \xrightarrow[\mathbf{x},\mathbf{y}]{k} \mathbf{y}\right) \le (\beta \cdot 4C)^{k-1} \mathbb{P}_{\mathbf{x},\mathbf{y}}(\mathbf{x} \to \mathbf{y}),$$

where $\mathbf{x} \xrightarrow[\mathbf{x},\mathbf{y}]{k} \mathbf{y}$ denotes the event that x is connected to $\mathbf{y}$ by a direct path of length k where all intermediate vertices are younger than $\mathbf{x}$ and $\mathbf{y}$. Let now $\mathbf{x}_0, \mathbf{x}_1, \ldots, \mathbf{x}_k = \mathbf{y}$ be a given skeleton and let us write $\mathbf{x} \xrightarrow[\mathbf{x}_0,\ldots,\mathbf{x}_k]{n} \mathbf{x}_k$ for the event that there is a directed path from $\mathbf{x}_0$ to $\mathbf{x}_k$ with skeleton $\mathbf{x}_0, \mathbf{x}_1, \ldots, \mathbf{x}_k$. We can then use the BK-inequality [5] in a version of [25] as outlined in [22, Eq.(11)] to deduce

$$\mathbb{P}_{\mathbf{x}_0,\mathbf{x}_1,\ldots,\mathbf{x}_k}\left(\mathbf{x} \xrightarrow[\mathbf{x}_0,\ldots,\mathbf{x}_k]{n} \mathbf{y}\right) \le (\beta C)^{n-k}\binom{n}{k}\prod_{j=1}^{k}\mathbb{P}_{\mathbf{x}_{j-1},\mathbf{x}_j}(\mathbf{x}_{j-1} \to \mathbf{x}_j) \qquad (4)$$

for some new constant $C > 1$. Define now A_n to be the event that $\mathbf{o}$ starts a directed path of length n that ends in the oldest vertex of the path. In other words, the path has a skeleton with strictly decreasing birth times. Using (4) and Mecke's equation [31], we obtain

$$\mathbb{P}(A_n) \leq \sum_{k=1}^{n} \left[(\beta C)^{n-k} \binom{n}{k} \int_{\substack{x_0=0, x_1, \dots x_k \in \mathbb{R}^d \\ 1>t_0>t_1>\dots>t_k}} \bigotimes_{j=0}^{k} d\mathbf{x}_j \prod_{j=1}^{k} \mathbb{P}_{\mathbf{x}_{j-1}, \mathbf{x}_j}(\mathbf{x}_{j-1} \to \mathbf{x}_j) \right]$$

The last equation can be easily calculated similarly as done in [22, Lemma 2.4] from which we infer $\mathbb{P}_0(A_n) \leq (\beta C)^n$ for some $C > 1$. Choosing $\beta < 1/C$, we infer from the Borel-Cantelli Lemma that almost surely there is a finite N such that A_n does not occur for all $n > N$. This however implies that every infinite path has bounded from below birth time because a path with birth times approaching zero contains sub paths ending in its oldest vertex of arbitrary length. This concludes the proof as it is easy to see that no such infinite paths can exist for small intensities β.

References

1. Aldous, D., Lyons, R.: Processes on unimodular random networks. Electron. J. Probab. **12**(54), 1454–1508 (2007). https://doi.org/10.1214/EJP.v12-463
2. Aldous, D., Steele, J.M.: The objective method: probabilistic combinatorial optimization and local weak convergence. In: Kesten, H. (eds.) Probability on discrete structures, Encyclopaedia Math. Sci., vol. 110, pp. 1–72. Springer, Heidelberg (2004). https://doi.org/10.1007/978-3-662-09444-0_1
3. Barabási, A.L., Albert, R.: Emergence of scaling in random networks. Science **286**(5439), 509–512 (1999). https://doi.org/10.1126/science.286.5439.509
4. Benjamini, I., Schramm, O.: Recurrence of distributional limits of finite planar graphs. Electron. J. Probab. **6**(23), 13 (2001). https://doi.org/10.1214/EJP.v6-96
5. van den Berg, J., Kesten, H.: Inequalities with applications to percolation and reliability. J. Appl. Probab. **22**(3), 556–569 (1985). https://doi.org/10.1017/s0021900200029326
6. Bloznelis, M., Götze, F., Jaworski, J.: Birth of a strongly connected giant in an inhomogeneous random digraph. J. Appl. Probab. **49**(3), 601–611 (2012). https://doi.org/10.1239/jap/1346955320
7. Bollobás, B., Riordan, O.: Robustness and vulnerability of scale-free random graphs. Internet Math. **1**(1), 1–35 (2003). https://doi.org/10.1080/15427951.2004.10129080
8. Bollobás, B., Riordan, O., Spencer, J., Tusnády, G.: The degree sequence of a scale-free random graph process. Random Struct. Alg. **18**(3), 279–290 (2001). https://doi.org/10.1002/rsa.1009
9. Broadbent, S.R., Hammersley, J.M.: Percolation processes. I. Crystals and mazes. Proc. Cambridge Philos. Soc. **53**, 629–641 (1957). https://doi.org/10.1017/s0305004100032680
10. Broder, A., et al.: Graph structure in the web. Comput. Netw. **33**(1), 309–320 (2000)

11. Burton, R.M., Keane, M.: Density and uniqueness in percolation. Comm. Math. Phys. **121**(3), 501–505 (1989). http://projecteuclid.org/euclid.cmp/1104178143
12. Cao, J., Olvera-Cravioto, M.: Connectivity of a general class of inhomogeneous random digraphs. Random Struct. Alg. **56**(3), 722–774 (2020). https://doi.org/10.1002/rsa.20892
13. Chen, N., Olvera-Cravioto, M.: Directed random graphs with given degree distributions. Stoch. Syst. **3**(1), 147–186 (2013). https://doi.org/10.1214/12-SSY076
14. Cirkovic, D., Wang, T., Resnick, S.I.: Preferential attachment with reciprocity: properties and estimation. J. Complex Netw. **11**(5), Paper No. cnad031, 41 (2023). https://doi.org/10.1093/comnet/cnad031
15. Daley, D.J., Vere-Jones, D.: An introduction to the theory of point processes. Vol. II. Probability and its Applications (New York), 2nd edn. General Theory and Structure. Springer, New York (2008). https://doi.org/10.1007/978-0-387-49835-5
16. Deijfen, M., van der Hofstad, R., Hooghiemstra, G.: Scale-free percolation. Ann. Inst. Henri Poincaré Probab. Stat. **49**(3), 817–838 (2013). https://doi.org/10.1214/12-AIHP480
17. Deprez, P., Wüthrich, M.V.: Construction of directed assortative configuration graphs. Internet Math. **1**(1) (2017). https://doi.org/10.24166/im.05.2017
18. Deprez, P., Wüthrich, M.V.: Scale-free percolation in continuum space. Commun. Math. Stat. **7**(3), 269–308 (2019). https://doi.org/10.1007/s40304-018-0142-0
19. Dereich, S., Mönch, C., Mörters, P.: Typical distances in ultrasmall random networks. Adv. Appl. Probab. **44**(2), 583–601 (2012). https://doi.org/10.1239/aap/1339878725
20. Dereich, S., Mörters, P.: Random networks with sublinear preferential attachment: degree evolutions. Electron. J. Probab. **14**(43), 1222–1267 (2009). https://doi.org/10.1214/EJP.v14-647
21. Gracar, P., Grauer, A., Lüchtrath, L., Mörters, P.: The age-dependent random connection model. Queueing Syst. **93**(3–4), 309–331 (2019). https://doi.org/10.1007/s11134-019-09625-y
22. Gracar, P., Lüchtrath, L., Mörters, P.: Percolation phase transition in weight-dependent random connection models. Adv. Appl. Probab. **53**(4), 1090–1114 (2021). https://doi.org/10.1017/apr.2021.13
23. Gracar, P., Lüchtrath, L., Mönch, C.: Finiteness of the percolation threshold for inhomogeneous long-range models in one dimension (2022)
24. Grimmett, G.: Percolation. In: Grundlehren der mathematischen Wissenschaften [Fundamental Principles of Mathematical Sciences], vol. 321, 2nd edn. Springer, Heidelberg (1999). https://doi.org/10.1007/978-3-662-03981-6
25. Heydenreich, M., van der Hofstad, R., Last, G., Matzke, K.: Lace expansion and mean-field behavior for the random connection model (2022)
26. van der Hofstad, R.: Random Graphs and Complex Networks. vol. 1, Cambridge Series in Statistical and Probabilistic Mathematics, vol. 43. Cambridge University Press, Cambridge (2017). https://doi.org/10.1017/9781316779422
27. van der Hofstad, R.: The giant in random graphs is almost local (2023)
28. van der Hofstad, R., van der Hoorn, P., Maitra, N.: Local limits of spatial inhomogeneous random graphs. Adv. Appl. Probab. **55**(3), 793–840 (2023). https://doi.org/10.1017/apr.2022.61
29. Jacob, E., Mörters, P.: Spatial preferential attachment networks: power laws and clustering coefficients. Ann. Appl. Probab. **25**(2), 632–662 (2015). https://doi.org/10.1214/14-AAP1006
30. Jacob, E., Mörters, P.: Robustness of scale-free spatial networks. Ann. Probab. **45**(3), 1680–1722 (2017). https://doi.org/10.1214/16-AOP1098

31. Last, G., Penrose, M.: Lectures on the Poisson Process. Cambridge University Press (2017). https://doi.org/10.1017/9781316104477
32. Meester, R., Roy, R.: Continuum percolation, Cambridge Tracts in Mathematics, vol. 119. Cambridge University Press, Cambridge (1996). https://doi.org/10.1017/CBO9780511895357
33. Mönch, C., Rizk, A.: Directed acyclic graph-type distributed ledgers via young-age preferential attachment. Stoch. Syst. **13**(3), 377–397 (2023). https://doi.org/10.1287/stsy.2022.0005
34. Penrose, M.D.: The strong giant in a random digraph. J. Appl. Probab. **53**(1), 57–70 (2016). https://doi.org/10.1017/jpr.2015.8
35. Trolliet, T., Cohen, N., Giroire, F., Hogie, L., Pérennes, S.: Interest clustering coefficient: a new metric for directed networks like Twitter. J. Complex Netw. **10**(1), Paper No. 30 (2022). https://doi.org/10.1093/comnet/cnab030

How to Cool a Graph

Anthony Bonato[✉], Holden Milne, Trent G. Marbach, and Teddy Mishura

Toronto Metropolitan University, Toronto, Canada
abonato@torontomu.ca

Abstract. We introduce a new graph parameter called the cooling number, inspired by the spread of influence in networks and its predecessor, the burning number. The cooling number measures the speed of a slow-moving contagion in a graph; the lower the cooling number, the faster the contagion spreads. We provide tight bounds on the cooling number via a graph's order and diameter. Using isoperimetric results, we derive the cooling number of Cartesian grids. The cooling number is studied in graphs generated by the Iterated Local Transitivity model for social networks. We conclude with open problems.

1 Introduction

The spread of influence has been studied since the early days of modern network science; see [14–16,18]. From the spread of memes and disinformation in social networks like X, TikTok, and Instagram to the spread of viruses such as COVID-19 and influenza in human contact networks, the spread of influence is a central topic. Common features of most influence spreading models include nodes infecting their neighbors, with the spread governed by various deterministic or stochastic rules. The simplest form of influence spreading is for a node to infect all of its neighbors.

Inspired by a desire for a simplified, deterministic model for influence spreading and by pursuit evasion games and processes such as Firefighter played on graphs, in [11,12], burning was introduced. Unknown to the authors then, a similar problem was studied in the context of hypercubes much earlier by Noga Alon [1]. In burning, nodes are either burning or not burning, with all nodes initially labeled as not burning. Burning plays out in discrete rounds or time-steps; we choose one node to burn in the first round. In subsequent rounds, neighbors of burning nodes themselves become burning, and in each round, we choose an additional source of burning from the nodes that are not burned (if such a node exists). Those sources taken in order of selection form a *burning sequence*. Burning ends on a graph when all nodes are burning, and the minimum length of a burning sequence is the *burning number* of G, denoted $b(G)$.

Much is now known about the burning number of a graph. For example, burning a graph reduces to burning a spanning tree. The burning problem is **NP**-complete on many graph families, such as the disjoint union of paths, caterpillars

Research supported by a grant of the first author from NSERC.

M. Dewar et al. (Eds.): WAW 2024, LNCS 14671, pp. 115–129, 2024.
https://doi.org/10.1007/978-3-031-59205-8_8

with maximum degree three, and spiders, which are trees where there is a unique node of degree at least three. For a connected graph G of order n, it is known that $b(G) \leq \sqrt{4n/3}+1$, and it is conjectured that the bound can be improved to $\lceil\sqrt{n}\rceil$ (which is the burning number of a path of order n). For more on burning, see the survey [4] and book [5].

Burning models explosive spread in networks, as in the case of a rapidly spreading meme on social media. However, as is the case for viral outbreaks such as COVID-19, while the spread may happen rapidly based on close contact, it can be mitigated by social distancing and other measures such as ventilation and vaccination. In a certain sense, a dual problem to burning is how to slow an infection as much as possible, as one would attempt to do in a pandemic. For this, we introduce a new contact process called cooling. Cooling spreads analogous to burning, with cooled neighbors spreading their infection to neighboring nodes, and a new cooling source is chosen in each round. However, where burning seeks to minimize the number of rounds to burn all nodes, cooling seeks to *maximize* the number of rounds to cool all nodes. Hence, burning and cooling are identical processes on graphs but with different objective functions.

More formally, given a finite, simple, undirected graph G, the cooling process on G is a discrete-time process. Nodes may be either *uncooled* or *cooled* throughout the process. Initially, in the first round, all nodes are uncooled. At each round $t \geq 1$, one new uncooled node is chosen to cool if such a node is available. We call such a chosen node a *source*. If a node is *cooled*, then it remains in that state until the end of the process. Once a node is cooled in round t, in round $t + 1$, its uncooled neighbors become cooled. A source is chosen in round $t + 1$ after the cooling spreads. The process ends in a given round when all nodes of G are cooled. Sources are chosen in each round for which they are available.

We define the *cooling number* of G, written $\mathrm{CL}(G)$, to be the maximum number of rounds for the cooling process to end. A *cooling sequence* is the set of sources taken in order during cooling. We have that $b(G) \leq \mathrm{CL}(G)$, and in some cases, the cooling number is much larger than the burning number. Note that while a choice of sources that burns the graph gives an upper bound to the burning number, a choice of sources that cools the graph gives a lower bound to the cooling number.

Consider a cycle C_8 with eight nodes for an elementary example of cooling. By symmetry, we may choose any node as the initial source. See Fig. 1 for a cooling sequence of length four. There is no longer cooling sequence, and so $\mathrm{CL}(C_8) = 4$.

The present paper aims to introduce cooling, provide bounds, and consider its value on various graph families. In Sect. 2, we will discuss cooling's relation to well-known graph parameters that provide various bounds on the cooling number. Using isoperimetric properties, we determine the cooling number of Cartesian grid graphs in the next section. The Iterated Local Transitivity (ILT) model, introduced in [10] and further studied in [2,6–8], simulates structural properties in complex networks emerging from transitivity. The ILT model simulates many properties of social networks. For example, as shown in [10], graphs

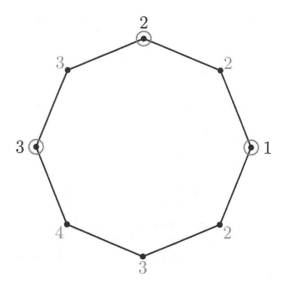

Fig. 1. An example of cooling on the cycle of length 8. Black labels indicate the nodes of the cooling sequence in increasing order. Blue labels indicate the round that the corresponding node was cooled. (Color figure online)

generated by the model densify over time, have small diameter, high local clustering, and exhibit bad spectral expansion. We derive results for cooling ILT graphs in Sect. 4 and prove that the cooling number of ILT graphs is dependent on the cooling number after two time-steps of the model. We finish with open problems on the cooling number.

All graphs we consider are finite, simple, and undirected. For further background on graph theory, see [19].

2   Bounds on the Cooling Number

As a warm-up, we consider bounds on the cooling number in terms of various graph parameters. We only consider connected graphs throughout the paper unless otherwise stated. We apply these to derive the cooling number of various graph families, such as paths, cycles, and certain caterpillars. The following elementary theorem bounds the cooling number via a graph's order.

Theorem 1. *For a graph G on n nodes, we have that*

$$\mathrm{CL}(G) \leq \left\lceil \frac{n+1}{2} \right\rceil.$$

Proof. One uncooled node is cooled from the cooling sequence during each round, except during the last round if all nodes are cooled. The cooling will spread to at least one additional uncooled node per round, except for the first round. This implies that there are

$$\left\lceil \frac{n+1}{2} \right\rceil + \left\lceil \frac{n+1}{2} \right\rceil - 1 \geq n$$

cooled nodes at the end of round $\left\lceil \frac{n+1}{2} \right\rceil$, and the result follows.   □

We next bound the cooling number by the diameter.

Theorem 2. *For a graph G, we have that*

$$\left\lceil \frac{\mathrm{diam}(G)+2}{2} \right\rceil \leq \mathrm{CL}(G) \leq \mathrm{diam}(G)+1.$$

Proof. For the upper bound, if some node v is cooled during the first round, then all nodes of distance at most i from v will be cooled by the end of round $i+1$. All nodes have distance at most $\mathrm{diam}(G)$ from v, so all nodes will be cooled by the end of round $\mathrm{diam}(G)+1$.

For the lower bound, we provide a cooling sequence that cools G in at least $\left\lceil \frac{\mathrm{diam}(G)+2}{2} \right\rceil$ rounds. Let $(v_0, v_1, \ldots, v_d)$ be a path of diameter length in G. The cooling sequence will be $\left(v_{2i-2} : 1 \leq i \leq \left\lceil \frac{\mathrm{diam}(G)+2}{2} \right\rceil \right)$.

Assume that at the start of round $i \geq 2$, each cooled node has distance at most $2i-4$ from v_0. After the cooling spreads in this round, every cooled node has distance at most $2i-3$ from v_0. As such, the node v_{2i-2} is uncooled and a possible choice for the next node in the cooling sequence. This node is now cooled, so starting round $i+1$, each cooled node has distance at most $2(i+1)-4$ from v_0. This sets up the recursion, where we note that the condition holds at the start of the second round. Since $2\left\lceil \frac{\mathrm{diam}(G)+2}{2} \right\rceil - 2 \leq \mathrm{diam}(G)+1$, the sequence provided is a cooling sequence.   □

We now apply these bounds to give that for the path P_n of order n,

$$\mathrm{CL}(P_n) = \left\lceil \frac{n+1}{2} \right\rceil = \left\lceil \frac{\mathrm{diam}(P_n)+2}{2} \right\rceil.$$

The upper bound follows from Theorem 1 and the lower bound from Theorem 2. In particular, the upper bound of Theorem 1 is tight. In passing, we note (with proof omitted) that for a cycle C_n, $\mathrm{CL}(C_n) = \left\lceil \frac{n+2}{3} \right\rceil$.

Using a more complicated example, we can show that the upper bound of Theorem 2 is also tight. Define the *complete caterpillar* of length d, CC_d, as the graph formed by appending one node to each non-leaf node of P_d. We call the nodes of P_d in CC_d the *spine* of the caterpillar. Note that CC_d has $n = 2d - 2$ nodes.

Theorem 3. *We have that*

$$\mathrm{CL}(\mathrm{CC}_d) = \left\lceil \frac{n+1}{2} \right\rceil = \mathrm{diam}(P_n)+1.$$

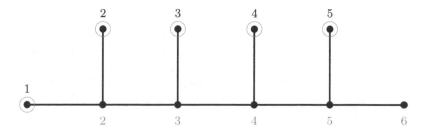

Fig. 2. An example of cooling on the complete caterpillar of length 6. Black labels indicate the nodes of the cooling sequence in increasing order. Blue labels indicate the round that the corresponding node was cooled. (Color figure online)

Proof. The upper bound follows from Theorem 1, and so to find the lower bound, we provide a strategy that takes $\lceil \frac{n+1}{2} \rceil$ rounds. Let $(v_1, v_2, \ldots, v_d)$ be the spine of the caterpillar, and let v_i' be the additional node that is appended to v_i, for $2 \leq i \leq d-1$. The cooling sequence will be v_1 followed by $(v_i' : 2 \leq i \leq d-1)$.

On round 1, the first node in the cooling sequence v_1 is cooled. At the start of round $i \geq 2$, the nodes $\{v_1, \ldots, v_{i-1}\} \cup \{v_2', \ldots, v_{i-1}'\}$ have been cooled. The cooling spreads, which cools only the node v_i. The next node in the cooling sequence is v_i'. At the start of round $i+1$, the nodes $\{v_1, \ldots, v_i\} \cup \{v_2', \ldots, v_i'\}$ have been cooled. This argument recursively repeats and ends on round d when v_d is cooled as cooling spreads. We then have that $\mathrm{CL}(\mathrm{CC}_d) = d = \lceil \frac{n+1}{2} \rceil$, as required. □

We finish the section by noting that determining the cooling number of certain graph families appears challenging. Even for spiders in general, determining the exact cooling number is not obvious. The following result provides bounds and exact values of the cooling numbers of certain spiders. A *spider* is a rooted tree that becomes a set of disjoint paths when the root node is removed; these paths are the spider's *legs*.

Theorem 4. *Let T be a spider with $2\,m$ legs, each of length r. If we have that $m < \lceil \log_2 r + 1 \rceil$, then*

$$\mathrm{CL}(T) \geq 2 \sum_{1 \leq i \leq m} \left\lfloor \frac{r+1}{2^i} \right\rfloor \sim (1 - 1/2^m)2r.$$

Otherwise,

$$\mathrm{CL}(T) = \mathrm{diam}(T) + 1.$$

Proof. We partition the cooling process into two major phases: before the head of the spider is cooled and after it is cooled. Let

$$m' = \min\{m, \lceil \log_2 r + 1 \rceil\}.$$

We further split each phase into m' subphases.

We start in the first phase. Suppose we enter subphase i, for i from 1 up to m'. This subphase runs for $\left\lceil \frac{r+1}{2^{(m'+1-i)}} \right\rceil$ rounds. In each round, we add the uncooled node in leg i furthest from the head to the cooling sequence. Note that after subphase i, each leg $i' < i$ will have at most $\frac{r+1}{2^{(m'-i)}} \leq r$ cooled nodes. At the end of this phase, there are m' legs with at most r cooled nodes, and the remaining at least m' legs do not have any cooled nodes. Note that the head of the spider has not yet been cooled.

The next phase begins. Suppose we enter subphase i of the second phase, for i from 1 up to m'. This subphase runs for $\left\lfloor \frac{r+1}{2^i} \right\rfloor$ rounds. In each round, we add the uncooled node in leg $m' + i$ to the cooling sequence that is closest to the head. Note that after subphase i, each leg $i' \leq m' + i$ can be assumed to be completely cooled, while legs $i' > m' + i$ will have $\sum_{1 \leq j \leq i} \left(\left\lfloor \frac{r+1}{2^j} \right\rfloor \right) - 1 < r$ cooled nodes (recalling that the cooling spreads at the start of a round).

A total of $2 \sum_{1 \leq i \leq m'} \left(\left\lfloor \frac{r+1}{2^i} \right\rfloor \right)$ rounds have been played, completing the result when $m < \lceil \log_2 r + 1 \rceil$. If $m \geq \lceil \log_2 r + 1 \rceil$, then note that the head was cooled in round $r + 1$ and that leg m did not contain a node in the cooling sequence. Therefore, the leaf of leg m was cooled on round $2r + 1 = \text{diam}(T) + 1$. □

3   Isoperimetric Results and Grids

This section studies cooling on Cartesian grids using isoperimetric results. Burning on Cartesian grids remains a difficult problem, with only bounds available in many cases. See [9, 17] for results on burning Cartesian grids.

Suppose G is a graph. For a $S \subseteq V(G)$, define its *node border*, $N(S)$, to be the set of nodes in $V(G) \setminus S$ that neighbor nodes in S. We then have the *node-isoperimetric parameter* of G at s as $\Phi_V(G, s) = \min_{S:|S|=s} |N(S)|$, and the *isoperimetric peak* of G as $\Phi_V(G) = \max_s \{\Phi_V(G, s)\}$. Note that in other work, it is common to use $\delta(S)$ to represent the node border in place of $N(S)$, but we use the chosen notation to prevent confusion with the minimum degree of a graph, $\delta(G)$.

Literature around node borders often either focuses on Cheeger's inequality, which is based on the ratio of $|N(S)|$ to $|S|$, or focuses on isoperimetric inequalities, which are bounds on $\Phi_V(G, s)$ from below. In recent work, for positive x and y, an inequality that bounds the maximum difference between the node-isoperimetric parameter at x and at $x + y$ was given while proving a distinct result (see Theorem 5 of [13]). We provide an analogous, reworded version of this bound with full proof for completeness.

Lemma 1 ([13]).   *For a graph G and integers $x, y \geq 0$,*

$$\Phi_V(G, x) - y \leq \Phi_V(G, x + y).$$

Proof. Let S_{x+y} be a set of nodes of cardinality $x + y$ with $|N(S_{x+y})| = \Phi_V(G, x + y)$. Note that if a node u is removed from S_{x+y}, then the only node that may be in the border of the new set $S_{x+y} \setminus \{u\}$ that was not in the border of S_{x+y} is the node u itself, as the nodes in the border are neighbors of nodes in the set. It follows that $N(S_{x+y} \setminus \{u\}) \subseteq N(S_{x+y}) \cup \{u\}$.

By a similar argument, if we remove any set $S_y \subseteq S_{x+y}$ of cardinality y from S_{x+y}, we have $N(S_{x+y} \setminus S_y) \subseteq N(S_{x+y}) \cup S_y$. This gives $|N(S_{x+y} \setminus S_y)| \leq |N(S_{x+y})| + y$, which re-arranges to yield

$$|N(S_{x+y} \setminus S_y)| - y \leq |N(S_{x+y})| = \Phi_V(G, x + y).$$

This completes the proof since $\Phi_V(G, x) \leq |N(S_{x+y} \setminus S_y)|$. $\qquad\square$

Lemma 1 can be considered a relative isoperimetric inequality. Define $x_1 = 1$, and let

$$x_{i+1} = x_i + \Phi_V(G, x_i) + 1.$$

In the following theorem, we will see that if S_i is the set of cooled nodes at time i, then $|S_i| \geq x_i$, and Lemma 1 must be used to prove this as it guarantees that the rate of cooling growth of a set of cooled vertices S_i' with $|S_i'| > x_i$ will always be high enough so that $|S_{i+1}'| \geq x_i + 1$.

This sequence of values x_i derives a natural upper bound on $\mathrm{CL}(G)$ as follows. Suppose I is the smallest value with $x_I \geq |V(G)|$.

Theorem 5. *If G is a graph, then we have that*

$$\mathrm{CL}(G) \leq I.$$

Proof. Suppose for the sake of contradiction that $\mathrm{CL}(G) \geq I + 1$. For some optimal cooling strategy, let S_j be the set of cooled nodes at the end of round j. We then have that $|S_1| = 1$. Assume that there is some round J where $|S_{J-1}| \geq x_{J-1}$ but $|S_J| < x_J$. Define $y = |S_{J-1}| - x_{J-1} \geq 0$. It follows by the definition of the cooling process, the definition of Φ_V, and by Lemma 1 that

$$
\begin{aligned}
|S_J| &= |S_{J-1}| + |N(S_{J-1})| + 1 \\
&\geq x_{J-1} + y + \Phi_V(G, x_{J-1} + y) + 1 \\
&\geq x_{J-1} + \Phi_V(G, x_{J-1}) - y + y + 1 \\
&= x_{J-1} + \Phi_V(G, x_{J-1}) + 1 = x_J.
\end{aligned}
$$

This contradicts the fact that $|S_J| < x_J$, and so we are done. $\qquad\square$

The (Cartesian) grid graph of length n, written G_n, is the graph with nodes of the form (u_1, u_2), where $1 \leq u_1, u_2 \leq n$, and an edge between (u_1, u_2) and (v_1, v_2) if $u_1 = v_1$ and $|u_2 - v_2| = 1$, or if $u_2 = v_2$ and $|u_1 - v_1| = 1$. The following total ordering of the nodes in the grid called the *simplicial ordering*, can be found in [3]. For two nodes $u = (u_1, u_2)$ and $v = (v_1, v_2)$, define $u < v$ when either $u_1 + u_2 < v_1 + v_2$, or $u_1 + u_2 = v_1 + v_2$ and $u_1 < v_1$. Let S_i be the i smallest nodes under this total ordering.

Lemma 2 ([3]). *No i-subset of $V(G)$ has a smaller node border than S_i, and so $|N(S_i)| = \Phi_V(G, i)$.*

Under the simplicial ordering, adding the smallest node in $V(G) \setminus S_i$ to S_i yields the set S_{i+1}. The set of nodes $N(S_i)$ is the set containing the $|\Phi_V(G, i)|$ smallest nodes that are larger than the nodes in S_i. As such, $S_i \cup N(S_i) = S_{i+\Phi_V(G,i)}$ and it also follows that $|S_i \cup N(S_i)| = i + \Phi_V(G, i)$.

Theorem 6. *An optimal cooling strategy for a grid graph G_n is formed by choosing the next node in a cooling sequence to be the smallest node in the simplicial ordering that has not yet been cooled.*

Proof. Let U_i be the set of nodes cooled by the end of round i when this strategy was performed. Note that for each i, there exists an x_i with $S_{x_i} = U_i$. We then have that $|U_i| = x_i$. During the next round, the cooling will spread to $N(U_i)$, meaning the nodes in $U_i \cup N(U_i) = S_{x_i + \Phi_V(G,x_i)}$ are now cooled.

We then cool the smallest uncooled node, implying that exactly the nodes $S_{x_i + \Phi_V(G,x_i)+1}$ are cooled, and so $U_{i+1} = S_{x_i + \Phi_V(G,x_i)+1}$. It follows that

$$x_{i+1} = |U_{i+1}| = x_i + \Phi_V(G, x_i) + 1.$$

If I is the smallest value with $x_I \geq |V(G)|$, then on all rounds $i < I$, $|U_i| < |V(G)|$, and so there is some uncooled node at the end of round i, and so play continues into round I. The process will, therefore, terminate under this approach exactly on round I. This proves that $\mathrm{CL}(G) \geq I$. Theorem 5 finishes the proof.   □

We explicitly determine the cooling number for Cartesian grids up to a small additive constant in the following result.

Theorem 7. *For each $n \geq 1$, there is an $\varepsilon \in \{0, 1, 2\}$ so that*

$$\mathrm{CL}(G_n) = 2n - 2\lfloor \log_2(n+3) \rfloor + \varepsilon.$$

Proof. We know the strategy of Theorem 6 is an optimal cooling strategy. In this particular strategy, the initial cooling sequence is straightforward to describe up until half of the vertices have been cooled, at which point the remaining vertices of the sequence become more complex. To prove this result, we consider the vertices that would be played by an optimal strategy up until almost half of the vertices would be cooled. This will be played over T rounds, and our analysis will find the first T vertices of the optimal cooling sequence. We then present an alternative strategy that uses these first T vertices of the optimal cooling strategy to construct a strategy that runs over $2T$ rounds; this shows that the optimal strategy must take at least $2T$ rounds. To conclude, we show that if there existed some strategy to cool the grid in $2T + 3$ rounds, then we would have a contradiction with the fact that an optimal strategy must cool almost half the vertices in T rounds.

We now begin by describing the optimal cooling strategy in terms of a number of phases. Phase 1 is just the first round, where only the node $(1, 1)$ is cooled. Phase p, with $2 \leq p \leq \lfloor \log_2(n+3) \rfloor$ starts with exactly those nodes of distance

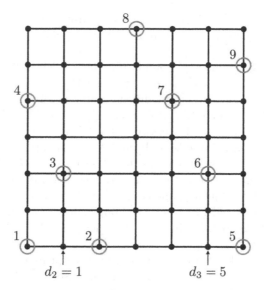

Fig. 3. An example of the cooling sequence for the 15×15 grid using the strategy of Theorem 7. The grid is a Cartesian product of paths, so vertex (u, v) is in row u and column v by convention. The nodes on the first row with distance d_2, d_3, d_4 from $(1, 1)$ are also indicated.

at most $d_p = 2^p - 4$ from $(1, 1)$ being cooled, and this phase will last for exactly $d_p + 3$ rounds.

In round t of phase p, with $1 \leq t \leq d_p + 3$, the cooling spreads to all uncooled nodes of distance $d_p + t$ from $(1, 1)$ and to the nodes $\{(t', d_p + t + 3 - t') : 1 \leq t' \leq 2t - 2\}$, which each have distance $d_p + t + 1$ from $(1, 1)$. The node $(2t - 1, d_p + 4 - t)$ is then chosen to be the next node of the cooling sequence, so the set of cooled nodes is exactly those nodes with distance at most $d_p + t$ from $(1, 1)$ and the nodes $\{(t', d_p + t + 3 - t') : 1 \leq t' \leq 2t - 1\}$, which have distance $d_p + t + 1$ from $(1, 1)$. At the end of round $d_p + 3$, this is exactly the set of nodes of distance $2d_p + 4 = 2^{p+1} - 4 = d_{p+1}$, and so we may iterate this procedure until phase $\lfloor \log_2(n + 3) \rfloor$.

Note that between the start and end of phase p, the ball of cooled nodes about $(1, 1)$ grows in radius by 2^p even though only $2^p - 1$ rounds have occurred. Thus, if T rounds have occurred from the start of play until some round during phase p, then the cooled nodes all sit within a ball of radius $T + p - 2$ around $(1, 1)$, and all the nodes in this ball will be cooled if the round is the last of phase p.

At the end of phase $p = \lfloor \log_2(n + 3) \rfloor - 1$, note that the cooled nodes form a ball of radius r around $(1, 1)$, where $\frac{n}{2} - 2 \leq r \leq n - 1$, and after phase p, the cooled nodes form a ball of radius r around $(1, 1)$, where $r \geq n$.

We will briefly discuss a different strategy that will help us bound the time the optimal cooling process will take. Suppose we have played the above strategy for the first $\lfloor \log_2(n+3) \rfloor - 1$ phases, and for phase $\lfloor \log_2(n+3) \rfloor$, we terminate the phase after the first round where a node of distance $n-2$ from $(1,1)$ was cooled.

Let T denote the number of rounds that have occurred by the end of this last modified phase. Note that $T + \lfloor \log_2(n+3) \rfloor - 2 = n-2$. We construct an additional T rounds for our modified strategy, which naturally split into $\lfloor \log_2(n+3) \rfloor$ phases. If node (r_i, c_i) was played on round i with $1 \leq i \leq T$, then we play node $(n+1-r_i, n+1-c_i)$ on round $2T+1-i$, for $1 \leq i \leq T$.

If phase p consisted of rounds t_1 through to t_2, then phase $2(\lfloor \log_2(n+3) \rfloor - 1) + 1 - p$ consists of rounds $2T+1-t_2$ through to $2T+1-t_1$. A similar analysis yields that each such node in the cooling sequence is uncooled before we cool it; hence, this is a cooling sequence, and cooling on G_n using this modified approach lasts for at least $2T$ rounds. If we proceed optimally, then the graph would take at least $2T$ rounds to cool. Noting that $T + \lfloor \log_2(n+3) \rfloor - 2 = n-2$, we have that $2T = 2n - 2\lfloor \log_2(n+3) \rfloor$.

Suppose, for the sake of contradiction, that the optimal approach lasts for at least $2T+3$ rounds. Note that after $T+3$ rounds, some nodes of distance $n+1$ from $(1,1)$ must have been cooled, and all nodes of distance n from $(1,1)$ must have been cooled

We now consider the reverse strategy, which first plays the last choice made in the optimal strategy, and so on. Note that by symmetry, after the first T rounds of this reverse strategy, the node that was chosen to be cooled has distance at least $n-2$ from (n,n). However, this means that the node chosen to be cooled on round T of this reverse strategy has distance at most n from $(1,1)$. Going back to the original optimal strategy, this means that round $T+4$ must have been played at a distance of at most n from $(1,1)$, but all of these nodes were cooled by round $T+3$. This gives us the desired contradiction, so the optimal strategy must last for at most $2T+2$ rounds. Therefore, the optimal strategy lasts for $2T$, $2T+1$, or $2T+2$ rounds, and the proof is complete. □

4   Cooling the ILT Model

Motivated by structural balance theory, the *Iterated Local Transitivity* (or *ILT* model) iteratively adds transitive triangles over time; see [10]. Graphs generated by these models exhibit several properties observed in complex networks, such as densification, small-world properties, and bad spectral expansion. The ILT model takes a graph $G = G_0$ as input and defines $\text{ILT}(G)$ by adding a cloned node x' for each node x in G, and making x' adjacent to the original node x and the neighbors of x. The set of cloned nodes is referred to as $\text{CLONE}(G)$, and we write $\text{ILT}_t(G)$ to denote the graph obtained by applying the ILT process t times to G. When $t = 1$, we typically write $\text{ILT}(G)$. We call any graph produced using at least one iteration of the ILT process an *ILT graph*. The diameter of $\text{ILT}(G)$ is the same as that of G unless the diameter of G is 1, in which case the diameter of $\text{ILT}(G)$ is 2.

The burning numbers of an ILT graph either equals $b(G)$ or $b(G)+1$; see [12]. Even though the order of the graphs generated by the ILT model grows

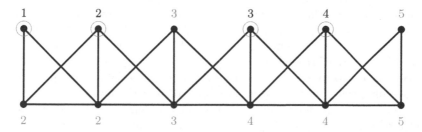

Fig. 4. An example of cooling on ILT(P_6). Black labels indicate the nodes of the cooling sequence in increasing order. Blue labels indicate the round that the corresponding node was cooled. (Color figure online)

exponentially with time, the burning number of ILT graphs remains constant. We now consider the cooling number of ILT graphs.

Theorem 8. *If G is a graph, then* $\mathrm{CL}(\mathrm{ILT}(G)) \geq \mathrm{CL}(G)$.

Proof. Let $(v_1, v_2, \ldots, v_k)$ be a maximum-length cooling sequence for G. Since the distance between v_i and v_j is the same in both G and in ILT(G), the sequence $(v_1, v_2, \ldots, v_k)$ is also a cooling sequence for ILT(G). If $\mathrm{CL}(G) = k$, then as the cooling process on ILT(G) with the given cooling sequence lasted k rounds, we have the result.

We may, therefore, assume that $\mathrm{CL}(G) = k + 1$ and that some node, say v, was one of the nodes cooled during the last round when cooling G. Since v must have a distance at least $k + 1 - i$ from v_i in G for this to occur, and since the distance from v to v_i is the same as this in ILT(G), it follows that v must have a distance at least $k + 1 - i$ from v_i in ILT(G). However, this implies that at the end of round k of the cooling process on ILT(G), node v has not yet been cooled. Thus, the cooling process on ILT(G) lasts at least $k + 1 = \mathrm{CL}(G)$ rounds, and so the proof is complete. $\square$

We next investigate the cooling number of the graphs $\mathrm{ILT}_t(P_n)$, where $t \geq 1$ and $P_n = (p_1, \ldots, p_n)$ is the path graph on n nodes. See Fig. 4 for an illustration of $\mathrm{ILT}_1(P_6)$. For $1 \leq k \leq t$, the k-th *layer set* of p_i is the set $L_k(p_i) = L_{k-1}(p_i) \cup \mathrm{CLONE}(L_{k-1}(p_i))$, where $L_0(p_i) = \{p_i\}$. When $k = t$, we write $L(p_i)$ instead.

Theorem 9. *If $n \geq 3$ and $t \geq 1$, then*

$$\mathrm{CL}(\mathrm{ILT}_t(P_n)) = \begin{cases} \lceil 2n/3 \rceil & \text{if } t = 1 \text{ and } n \equiv 2 \pmod 3; \\ \lceil 2n/3 \rceil + 1 & \text{otherwise.} \end{cases}$$

Proof. Let $G = \mathrm{ILT}_t(P_n)$ and let $(v_1, \ldots, v_k)$ be a cooling sequence for G. We begin with an upper bound on k. As the layer sets $\{L(p_i)\}$ form a partition of $V(G)$, each element v_i of the cooling sequence is in exactly one $L(p_j)$, for some $1 \leq j \leq n$. We claim that any three consecutive layer sets, those of the form $L(p_j), L(p_{j+1}), L(p_{j+2})$, can contain at most two sequence elements total.

Indeed, suppose that v, w are nodes belonging to any of these three layer sets. Since p_j and p_{j+2} are adjacent to p_{j+1} in P_n and by definition of the ILT process, v and w are adjacent to p_{j+1} in G. As such, $d(v, w) \leq 2$. It follows that if a node u in any of those layer sets were cooled, all three layer sets would be fully cooled in two rounds. In each round, the selection of a node in the cooling sequence happens after the cooling has spread from the nodes that were cooled by the end of the previous round, so only one node in the three layer sets could be selected after u was cooled, proving the claim. Thus, as there are n layer sets, we can separate them into $\lfloor n/3 \rfloor$ consecutive triples, each of which has at most two cooled members. If n is not divisible by 3, then the remaining layer sets may also each have a cooled element. This yields that $k \leq \lceil 2n/3 \rceil$.

This upper bound is also achieved, which we show by constructing a cooling sequence $(w_1, \ldots, w_k)$ of the desired length. Let the clone of the node $p_i \in V(\mathrm{ILT}_{t-1}(P_n))$ that was created in the t-th application of the ILT process be denoted p_i'. For $1 \leq i \leq \lceil 2n/3 \rceil$, define $w_i = p_{i+\lfloor (i-1)/2 \rfloor}'$. Note that this is well defined since $i + \lfloor (i-1)/2 \rfloor \leq n$ for all such i. We claim this sequence is a cooling sequence. Indeed, for any $1 \leq i < j \leq n$ we have that $d(p_i', p_j') = \max(2, |i - j|)$, as the path $(p_i', p_{i+1}, p_{i+2}, \ldots, p_{j-1}, p_j')$ is of this length and any path between nodes in layer sets $L(p_i)$ and $L(p_j)$ must cross through every layer set $L(p_k)$ with $i < k < j$, implying no shorter path can exist. Thus, the sequence is a cooling sequence of length $\lceil 2n/3 \rceil$, and as this matches the upper bound, this cooling sequence is optimal.

We have shown that the length of an optimal cooling sequence of G is $\lceil 2n/3 \rceil$. However, it is possible that the cooling process could last one round longer. Note the following. After round i, p_j' has been cooled for $1 \leq j \leq i + \lfloor \frac{i-1}{2} \rfloor$. This occurs since we chose to cool p_j' as it was in the cooling sequence, unless $j \equiv 0 \pmod 3$, in which case p_{j-2}' was chosen to be cooled as it was in the cooling sequence, and then two turns have occurred, causing p_j' to be cooled. After round i, we have that p_j is cooled for $1 \leq j \leq i + \lfloor \frac{i-1}{2} \rfloor - 1$, since p_{j-1}' was cooled in a round before round i, and then the cooling propagated. If $i \equiv 0 \pmod 2$, then p_j is also cooled for $j = i + \lfloor \frac{i-1}{2} \rfloor$, since then p_{j-1}' was cooled in round $i - 1$ and p_j is adjacent to p_{j-1}'. We now consider two cases.

Case 1: $t = 1$ and $n \equiv 2 \pmod 3$. In this case, $k \equiv 0 \pmod 2$, and so we have p_j and p_j' have been cooled by round k for $1 \leq j \leq k + \lfloor \frac{k-1}{2} \rfloor = n$. This is all the vertices in the graph, and so all vertices have been cooled by the end of round k.

Case 2: If $n \equiv 0, 1 \pmod 3$, then the vertex p_{n-2}' was cooled during round $k - 1$ due to being in the cooling sequence, and all vertices in $L(p_j)$ were cooled before this point for $1 \leq j \leq n - 3$. The cooled vertex p_{n-2}' caused vertex p_{k-1} to be cooled by propagation in round k, and in addition, all vertices in $L(p_{n-2})$ must be cooled during this propagation. However, no vertices in $L(p_n)$ have been cooled by the end of round k. In round k_1, all vertices in $L(p_n)$ are cooled by propagation, and so all vertices have been cooled.

If $n \equiv 2 \pmod 3$ and $t > 1$, then by the end of round $k - 1$, all vertices in $L(p_j)$ were cooled for $1 \leq j \leq n - 2$, and the vertex p_{n-1}' was just cooled due to being in the cooling sequence, but no other vertices have been cooled. In round k, the uncooled vertices in $L(p_{n-1})$ are cooled through propagation, as is some subset of vertices in $L_{t-1}(p_n)$. Additionally, vertex p'_n is cooled due to being in the cooling sequence. Since $t > 1$, $\mathrm{CLONE}(L_{t-1}(p_n)) \setminus \{p'_n\}$ is nonempty, and so there is at least one vertex in this set that is uncooled by the end of round k. All vertices in $L(p_n)$ are cooled in round $k+1$, and so all vertices are cooled. $\square$

The next result shows that the second time-step determines the cooling number of ILT graphs.

Theorem 10. *For any graph G, the maximum length of a cooling sequence in* $\mathrm{ILT}_2(G)$ *and* $\mathrm{ILT}_t(G)$ *are the same.*

Proof. Suppose that $(u_1, u_2, \ldots, u_d)$ is an optimal cooling sequence in $\mathrm{ILT}_t(G)$. Given a node u in $\mathrm{ILT}_t(G)$, let $f(u)$ be the node in G that was cloned to form u. Since $t \geq 2$, for each $x \in V(G)$, there are at least two distinct clone nodes in $\mathrm{ILT}_t(G)$ that were created in the latest iteration of the ILT process, say x' and x'', such that $f(x') = f(x'') = x$. Note that by the definition of the ILT model, if u is adjacent to v, then either $f(u)$ is adjacent to $f(v)$ or $f(u) = f(v)$.

We define a sequence of nodes in $\mathrm{ILT}_2(G)$ of length d, $(v_1, v_2, \ldots, v_d)$, as follows. If $f(u_i) \neq f(u_{i-1})$, then define $v_i = f(u_i)'$; otherwise, define $v_i = f(u_i)''$. We will show that this is an optimal cooling sequence for $\mathrm{ILT}_2(G)$.

To begin, we show that this is a valid cooling sequence. Assume for a contradiction that this sequence is not a valid cooling sequence. As such, there must be some v_j that will already be cool when we try to cool it on round j. For this to have occurred, there must some $i < j$ and some path from v_i to v_j of length at most $j - i$, say $(p_i, p_{i+1}, \ldots, p_j)$, such that node p_r was cooled on round r for each $r \in \{i, i+1, \ldots, j\}$. The walk $(f(p_i), f(p_{i+1}), \ldots, f(p_j))$ in $\mathrm{ILT}_2(G)$ has length $j - i$, and so contains a path of length $r' \leq j - i$. For simplicity, we may assume that $(f(p_i), f(p_{i+1}), \ldots, f(p_j))$ is a path in $\mathrm{ILT}_2(G)$. Since $f(p_i) = f(v_i) = f(u_i)$ and $f(p_j) = f(v_j) = f(u_j)$, it follows that $(u_i, f(p_{i+1}), \ldots, f(p_{j-1}), u_j)$ is a path of length $j - i$ in $\mathrm{ILT}_t(G)$. Since u_i was cooled in round i while cooling the graph $\mathrm{ILT}_t(G)$, and the cooling spreads to each uncooled neighbor over each round, it follows that u_j must have been cooled within $j - i$ rounds. We then have that u_j was cooled by round $i + (j - i) = j$, so u_j was already cooled by round j. However, $(u_1, u_2, \ldots, u_d)$ is a cooling sequence in $\mathrm{ILT}_t(G)$, and so we have the required contradiction.

To show that the cooling sequence $(v_1, \ldots, v_d)$ in $\mathrm{ILT}_2(G)$ is optimal, assume that there is some cooling sequence $(v_1, \ldots, v_{d+1})$ in $\mathrm{ILT}_2(G)$. This sequence of nodes is also a sequence of nodes in $\mathrm{ILT}_t(G)$, and further, the distance between any of these nodes is the same in both $\mathrm{ILT}_2(G)$ and $\mathrm{ILT}_t(G)$. In $\mathrm{ILT}_t(G)$, since no cooling sequence can have length $d + 1$, there must be a pair of nodes v_i and v_j, $i < j$, with $d(v_i, v_j) \leq j - i$. This follows by a similar argument to the first part of the proof. We then have that v_j will already be cooled by the time we try to cool it on turn j, contradicting that this is a cooling sequence. Thus, the maximum length of a cooling sequence in $\mathrm{ILT}_2(G)$ is d, and the proof follows. $\square$

If a graph G has a maximum cooling sequence of length s, then $\mathrm{CL}(G) \in \{s, s+1\}$. Theorem 8 assures us that applying the ILT process only increases the cooling value. We thus have the following result proving that, as in the case of burning, the cooling number remains bounded by a constant throughout the ILT process.

Corollary 1. *For any graph G, we have that*

$$\mathrm{CL}(\mathrm{ILT}_2(G)) \leq \mathrm{CL}(\mathrm{ILT}_t(G)) \leq \mathrm{CL}(\mathrm{ILT}_2(G)) + 1.$$

In particular, Theorem 8 and Corollary 1 together imply that for every graph G, either $\mathrm{CL}(\mathrm{ILT}_t(G)) = \mathrm{CL}(\mathrm{ILT}_2(G))$ for all $t \geq 2$, or there exists a threshold value t_0 such that the cooling number of $\mathrm{ILT}_t(G)$ is equal to $\mathrm{CL}(\mathrm{ILT}_2(G)) + 1$ if and only if $t \geq t_0$.

5   Conclusion and Further Directions

We introduced the cooling number of a graph, which quantifies the spread of a slow-moving contagion in a network. We gave tight bounds on the cooling number as functions of the order and diameter of the graph. Using isoperimetric techniques, we determined the cooling number of Cartesian grids. The cooling number of ILT graphs was considered in the previous section.

Several questions remain on the cooling number. Determining the exact value of the cooling number in various graph families, such as spiders and, more generally, trees, remains open. In the full version of the paper, we will consider the cooling number of other grids, such as strong or hexagonal grids. We want to classify the cooling number of ILT graphs where the initial graphs are not paths. Another direction to consider is the complexity of cooling, which is likely **NP**-hard.

References

1. Alon, N.: Transmitting in the n-dimensional cube. Discret. Appl. Math. **37**, 9–11 (1992)
2. Behague, N., Bonato, A., Huggan, M.A., Malik, R., Marbach, T.G.: The iterated local transitivity model for hypergraphs. Discret. Appl. Math. **337**, 106–119 (2023)
3. Bollobás, B., Leader, I.: Compressions and isoperimetric inequalities. J. Comb. Theory Ser. A **56**(1991), 47–62 (1991)
4. Bonato, A.: A survey of graph burning. Contrib. Discret. Math. **16**, 185–197 (2021)
5. Bonato, A.: An Invitation to Pursuit-Evasion Games and Graph Theory. American Mathematical Society, Providence, Rhode Island (2022)
6. Bonato, A., Chaudhary, K.: The iterated local transitivity model for tournaments. In: Proceedings of WAW 2023 (2023)
7. Bonato, A., Chuangpishit, H., English, S., Kay, B., Meger, E.: The iterated local model for social networks. Discret. Appl. Math. **284**, 555–571 (2020)

8. Bonato, A., Cranston, D.W., Huggan, M.A., Marbach, T.G., Mutharasan, R.: The iterated local directed transitivity model for social networks. In: Proceedings of WAW 2020 (2020)
9. Bonato, A., English, S., Kay, B., Moghbel, D.: Improved bounds for burning fence graphs. Graphs and Combinatorics **37**, 2761–2773 (2021)
10. Bonato, A., Hadi, N., Horn, P., Prałat, P., Wang, C.: Models of on-line social networks. Internet Math. **6**, 285–313 (2011)
11. Bonato, A., Janssen, J., Roshanbin, E.: Burning a graph as a model of social contagion. In: Proceedings of WAW 2014 (2014)
12. Bonato, A., Janssen, J., Roshanbin, E.: How to burn a graph. Internet Math. **1–2**, 85–100 (2016)
13. Bonato, A., Marbach, T., Marcoux, J., Nir, J.D.: The k-visibility Localization game, Preprint (2023)
14. Domingos, P., Richardson, M.: Mining the network value of customers. In: Proceedings of the 7th International Conference on Knowledge Discovery and Data Mining (KDD) (2001)
15. Kempe, D., Kleinberg, J., Tardos, E.: Maximizing the spread of influence through a social network, In: Proceedings of the 9th International Conference on Knowledge Discovery and Data Mining (KDD) (2003)
16. Kempe, D., Kleinberg, J., Tardos, E.: Influential nodes in a diffusion model for social networks. In: Proceedings 32nd International Colloquium on Automata, Languages and Programming (ICALP) (2005)
17. Mitsche, D., Prałat, P., Roshanbin, E.: Burning graphs-a probabilistic perspective. Graphs and Combinatorics **33**, 449–471 (2017)
18. Richardson, M., Domingos, P.: Mining knowledge-sharing sites for viral marketing. In: Proceedings of the 8th International Conference on Knowledge Scovery and Data Mining (KDD) (2002)
19. West, D.B.: Introduction to Graph Theory, 2nd edn. Prentice Hall (2001)

Distributed Averaging for Accuracy Prediction in Networked Systems

Christel Sirocchi$^{(\boxtimes)}$ and Alessandro Bogliolo

Department of Pure and Applied Sciences, University of Urbino, Piazza della Repubblica 13, 61029 Urbino, Italy
`c.sirocchi2@campus.uniurb.it`

Abstract. Distributed averaging is among the most relevant cooperative control problems, with applications in sensor and robotic networks, distributed signal processing, data fusion, and load balancing. Gossip algorithms have been investigated and successfully deployed in multi-agent systems to perform distributed averaging in asynchronous settings. This study proposes a heuristic approach to estimate the convergence rate of averaging algorithms in a distributed manner, relying on the computation and propagation of local graph metrics while entailing simple data elaboration and small message passing. The proposed strategy enables nodes to predict the number of interactions needed to estimate the global average with the desired accuracy. Consequently, nodes can make informed decisions on their use of measured and estimated data while gaining awareness of the global structure of the network. The study presents applications to outliers identification and performance evaluation in switching topologies.

Keywords: Distributed computing · Consensus · Graph metrics

1 Introduction

Distributed averaging is an instance of distributed computation aiming to determine the global average of a set of values by iterating local calculations. It has been extensively studied as the primary tool for solving cooperative problems in multi-agent systems such as sensor networks, micro-grids, transport networks, and power distribution systems [12]. In these settings, the value of an aggregate function over the entire network data is often more relevant than individual data at nodes. Networks of temperature sensors, for instance, are generally deployed to assess the average temperature in a given area. Similarly, peer-to-peer systems are primarily interested in calculating the average size of stored files [10]. Other collective behaviours leveraging distributed averaging include formation control of autonomous vehicles, network synchronisation, and automated traffic networks [24]. Recent studies focused on distributed averaging as a central subroutine in large-scale optimisation problems, specifically in distributed deep learning systems, where training a model on a massive dataset can be expedited by performing mini-batch gradient updates across multiple machines and synchronising learning parameters by distributed averaging [1].

M. Dewar et al. (Eds.): WAW 2024, LNCS 14671, pp. 130–145, 2024.
https://doi.org/10.1007/978-3-031-59205-8_9

The performance of a distributed averaging protocol is generally evaluated as the communication rounds required by each node to obtain an estimate of the global average with the desired level of accuracy. Performance analysis is fundamental in networked systems, as they typically suffer limitations in terms of communication bandwidth, memory, and computational power [22]. Seminal work on gossip algorithms derived theoretical performance guarantees depending on the eigenvalues of the matrix characterising the algorithm [2]. However, these results cannot help nodes make local and accurate performance predictions, as they identify wide performance intervals and require knowledge of the entire network structure for eigenvalues calculations.

In networks where nodes lack awareness of the global network structure and can only communicate with neighbours, local graph metrics and their summative statistics computed distributively (e.g., by distributed averaging) can offer nodes a glimpse into the network's global properties and guide nodes in adding, removing or rewiring some of their connections to increase performance. For instance, the clustering coefficient estimates propagation of redundant information [4] and local efficiency quantifies the impact of node loss in communications [19], while the average node degree determines the overall level of cooperation and robustness in a communication network [20].

This work examines the predictive power of averages of local metrics over the performance of a distributed averaging gossip algorithm, intended as the number of transmissions or time each node requires to estimate the global average with the desired level of accuracy. Consequently, it introduces a novel method enabling nodes to predict the performance of the averaging algorithm in a distributed manner, entailing simple data elaboration and small message passing. As a result, agents can perform the necessary communication rounds to achieve the desired accuracy, make informed decisions on their use of measured and estimated data and also gain awareness of the global network structure.

The remainder of the paper is organised as follows. Section 2 provides some relevant background on averaging algorithms, citing previous results on convergence in asynchronous models, while Sect. 3 outlines the proposed approach. Section 4 presents relevant applications, and finally, Sect. 5 provides conclusions and directions for future work.

2   Background

2.1   Network Topology

The communication constraints in networked systems can be conveniently modelled via a graph $G = (V, E)$, where V is the vertex set of n nodes v_i, with $i \in I = \{1, \ldots, n\}$ and $n \in \mathbb{N}$, and E is the edge set $E \subseteq V \times V$ of the pairs $e_{ij} = (v_i, v_j)$, so that there is an edge between nodes v_i and v_j iff $(v_i, v_j) \in E$. All nodes that can transmit information to node v_i are said to be its neighbours and are represented by the set $\Omega_i = \{v_j : (v_i, v_j) \in E\}$. In undirected graphs, edges $e_{ij} \in E$ are unordered pairs $\{v_i, v_j\}$ of elements of V, reflecting bidirectional communication among nodes.

2.2 Distributed Average

Let x_i denote the value of node v_i, representing a physical quantity such as position, temperature, light intensity or voltage, and $\mathbf{x} = (x_1, ..., x_n)^T$ the vector of values so that the ith component of $\mathbf{x}$ is the value at node v_i. The nodes v_i and v_j are said to agree in a network iff $x_i = x_j$. All nodes in G are in agreement or have reached a consensus iff $x_i = x_j \, \forall i, j \in I$. This agreement space can also be expressed as $\mathbf{x} = \alpha \mathbf{1}$ where $\mathbf{1} = (1, ..., 1)^T$ and $\alpha \in R$ is the collective decision value of the nodes. The system is said to reach asymptotic consensus if all nodes asymptotically converge to α [5], i.e.

$$\lim_{t \to +\infty} x_i(t) = \alpha, \forall i \in I.$$

A consensus algorithm (or protocol) is an interaction rule that specifies the information exchange between an agent and its neighbours to reach a consensus [14], i.e. to asymptotically converge to the agreement space [23]. One of the benefits of using linear iteration-based schemes is that each node only transmits a single value to each of its neighbours [23]. In discrete time, the consensus protocol is

$$\mathbf{x}(k) = \mathbf{W}(k) \, \mathbf{x}(k-1) \tag{1}$$

where $\mathbf{x}(k)$ is the vector of values at the end of time slot k, and $\mathbf{W}$ is the $n \times n$ matrix of the averaging weights. When the consensus value corresponds to the average of all initial values, i.e. $\alpha = \frac{1}{n} \sum_{i=1}^{n} x_i(0)$, the system is said to perform distributed averaging. In consensus protocols with time-invariant weight matrix $\mathbf{W}$, the linear iteration implies

$$\mathbf{x}(k) = \mathbf{W}^k \, \mathbf{x}(0). \tag{2}$$

To achieve asymptotic average consensus regardless of the initial values $\mathbf{x}(0)$

$$\lim_{t \to +\infty} \mathbf{W}^t = \frac{\mathbf{1}\mathbf{1}^T}{n} \tag{3}$$

which follows from

$$\lim_{t \to +\infty} \mathbf{x}(t) = \lim_{t \to +\infty} \mathbf{W}^t \mathbf{x}(0) = \frac{\mathbf{1}\mathbf{1}^T}{n} \mathbf{x}(0). \tag{4}$$

Equation 3 holds iff the following three properties are satisfied:

$$\mathbf{1}^T \mathbf{W} = \mathbf{1}^T \tag{5}$$

i.e. $\mathbf{1}$ is a left eigenvector of $\mathbf{W}$ associated with the eigenvalue 1, implying that $\mathbf{1}^T \mathbf{x}(k+1) = \mathbf{1}^T \mathbf{x}(k) \, \forall \, k$, i.e., the sum, and therefore the average, of the vector of node values is preserved at each step;

$$\mathbf{W} \, \mathbf{1} = \mathbf{1} \tag{6}$$

i.e. $\mathbf{1}$ is also a right eigenvector of $\mathbf{W}$ associated with the eigenvalue 1, meaning that $\mathbf{1}$, or any vector with constant entries, is a fixed point for the linear iteration;

$$\rho(\mathbf{W} - \frac{\mathbf{1}\mathbf{1}^T}{n}) < 1 \tag{7}$$

where $\rho(\cdot)$ denotes the spectral radius of a matrix, which, combined with the first two conditions, states that 1 is a simple eigenvalue of $\mathbf{W}$, and all other eigenvalues have a magnitude strictly less than 1 [26].

The disagreement vector $\boldsymbol{\delta}(k)$ quantifies the distance from consensus at time slot k and can be computed as

$$\boldsymbol{\delta}(k) = \mathbf{x}(k) - \alpha\mathbf{1}. \tag{8}$$

Notably, this vector evolves according to the same linear system as the vector:

$$\boldsymbol{\delta}(k) = \mathbf{W}(k)\boldsymbol{\delta}(k-1). \tag{9}$$

Iterative distributed averaging algorithms are often classified based on the adopted time model. Gossip algorithms realise asynchronous averaging schemes so that a single pair of neighbouring nodes interact at each time k, setting their values to the average of their previous values [3]. In contrast, consensus algorithms implement a synchronous model where time is commonly slotted across nodes, and all nodes simultaneously update their values with a linear combination of the values of their neighbours at discrete times k. Gossip protocols are more suited to model real networks and are generally more accessible to implement for the lack of synchronisation requirements, unrealistic for most applications [15], but are harder to characterise mathematically due to the added randomness of the neighbour selection.

2.3 Gossip Algorithms

In asynchronous gossip protocols, a single node is active at each time slot k and selects one of its neighbours for interaction according to a given criterion. The $n \times n$ probability matrix $\mathbf{P} = [p_{ij}]$ prescribes the probability p_{ij} that the node v_i selects node v_j, with $p_{ij} = 0$ if $(v_i, v_j) \notin E$ due to the constraints of only interacting with neighbours. For instance, in random neighbour selection, where all neighbours are equally likely to be chosen, the matrix $\mathbf{P}$ is $[p_{ij}]$ such that $p_{ij} = 1/|\Omega_i| \; \forall v_j \in \Omega_i$ and 0 otherwise. A node v_i interacts with node v_j at time slot k with probability $\frac{p_{ij}}{n}$, which is the joint probability that v_i is active at time slot k $(p = \frac{1}{n})$ and selects node v_j for interaction $(p = p_{ij})$. The weight matrix $\mathbf{W}_{ij}$ of this averaging scheme has elements

$$w_{kl}(t) = \begin{cases} \frac{1}{2} & \text{if } k, l \in \{i, j\} \\ 1 & \text{if } k = l, k \notin \{i, j\} \\ 0 & \text{otherwise.} \end{cases} \tag{10}$$

This is equivalent to nodes v_i and v_j setting their values to the average of their current values, leaving the others unchanged. The matrix $\mathbf{W}$ generally changes over time, as different pairs interact at each time slot. The averaging process is thus defined by the sequence of averaging matrices $\{\mathbf{W}(k)\}_k$ and the vector value at time step k can be computed as

$$\mathbf{x}(k) = \mathbf{W}(k-1)\mathbf{W}(k-2) \ .. \ \mathbf{W}(0)\mathbf{x}(0) = \phi(k-1)\mathbf{x}(0) \tag{11}$$

Recalling that for independent real-valued random matrices the expected matrix of the product is the product of the expected matrices, then

$$\mathbb{E}(\phi(k)) = \prod_{i=0}^{k} \mathbb{E}(\mathbf{W}(i)) = \bar{\mathbf{W}}^k \tag{12}$$

where $\bar{\mathbf{W}}$ is the expected weight matrix

$$\bar{\mathbf{W}} = \sum_{i,j} \frac{p_{ij}}{n} \mathbf{W}_{ij} \tag{13}$$

most commonly written as

$$\bar{\mathbf{W}} = \mathbf{I} - \frac{1}{2n}\mathbf{D} + \frac{\mathbf{P}+\mathbf{P}^T}{2n}, \tag{14}$$

[3] where $\mathbf{I}$ is the identity matrix and $\mathbf{D}$ is the diagonal matrix with entries

$$\mathbf{D}_i = \sum_{j=1}^{n} [p_{ij} + p_{ji}].$$

By definition, $\bar{\mathbf{W}}$ is symmetric and doubly stochastic. If the underlying graph is connected and non-bipartite, the expected matrix $\bar{\mathbf{W}}$ fulfils all three conditions for convergence (Eq. 5, 6, 7), so the sequence of averaging matrices $\{\mathbf{W}(k)\}_k$ drawn independently and uniformly and applied to any initial vector $\mathbf{x}(0)$, converges in expectation to the vector average $\frac{\mathbf{1}\mathbf{1}^T}{n}\mathbf{x}(0)$ [3].

Notably, the performance of the averaging gossip scheme is determined by the second largest eigenvalue of the matrix $\bar{\mathbf{W}}$, as demonstrated in a seminal work by Boyd et al. [2], which provides a tight characterisation of the averaging time contingent on this eigenvalue.

2.4   Convergence Rate and Accuracy

The convergence of the distributed iteration is governed by the product of matrices, each of which satisfies certain communication constraints imposed by the graph topology and the gossip criterion. Let $\boldsymbol{\delta}(t)$ be the collective disagreement vector of $\mathbf{x}(t)$ at time t, and $\hat{\delta}(t)$ the normalised collective disagreement relative to the initial values $\mathbf{x}(0)$:

$$\hat{\delta}(t) = \frac{\|\boldsymbol{\delta}(t)\|}{\|\mathbf{x}(0)\|}, \tag{15}$$

where $\|.\|$ is the l_2 norm of the vector. Notably, the l_2 norm of the disagreement vector $\boldsymbol{\delta}(t)$, by definition, corresponds to the standard deviation of the vector $\mathbf{x}(t)$ scaled by the squared root of the network size:

$$\sigma(t) = \frac{\|\boldsymbol{\delta}(t)\|}{\sqrt{n}}. \tag{16}$$

Numerical and theoretical results for asynchronous averaging schemes indicate that the logarithm of the collective disagreement $\hat{\delta}(t)$ decreases linearly after a faster transient phase, and that the decreasing rate is deterministic and independent of the initial measurements $\mathbf{x}(0)$ [6]. Hence, a contraction rate γ can be defined as the angular coefficient of the linear stationary regime and used to characterise the algorithm performance [21]:

$$\log(\hat{\delta}) = -\gamma\,t + z. \tag{17}$$

The accuracy R of the estimates at time t can be quantified as the reduction in collective disagreement compared to the initial state, defined as:

$$R(t) = -\log(\frac{\hat{\delta}(t)}{\hat{\delta}(0)}). \tag{18}$$

As per its definition, $R(t) = \gamma\,t$, thus the time required to attain the desired accuracy level can be computed as:

$$t = R/\gamma. \tag{19}$$

If nodes initiate, on average, one interaction per unit of time, the time parameter t approximates the number of interactions per node.

3   Proposed Approach

3.1   Problem Setup

The study focuses on networks of agents, each initialised with a value $x_i(0)$ representing an opinion or a measurement, and engaged in distributed averaging to reach consensus on the global average α. The system leverages an asynchronous gossip protocol for distributed averaging, as detailed in Sect. 2.3, where agents engage in random neighbour selection. In this proposed scenario, each agent v_i in the network is assigned a parameter R_i denoting the required level of accuracy for its estimated average $x_i(t)$ to be confidently utilised, processed, shared with a remote server, or compared against its initial measurement. This parameter R indicates that the agent can trust its estimate only if the collective disagreement in the network is less than the initial disagreement reduced by a factor of 10^R, as defined in Eq. 18. The choice of each R depends on the agent's function within the network; for example, sensor nodes might be satisfied with a substantial reduction in group disagreement (e.g., $R = 4$), whereas control nodes may require estimates of many orders of magnitude more accurate (e.g., $R = 8$). In the considered setting, nodes can only exchange information with their immediate neighbours without access to the overall graph structure. Nevertheless, nodes can utilise information exchanged with neighbours to compute local graph metrics, such as their degree, clustering coefficient, and local efficiency.

After confirming the predictive value of local metric averages over the algorithm convergence rate, the study recommends that nodes compute local metrics and estimate their average across the network by distributed averaging, exploiting the same principle used to calculate the average of measured quantities. Nodes then employ averaged local metrics to estimate the algorithm convergence rate and make predictions of the time or number of interactions needed to achieve the desired accuracy so that nodes use their estimate only when confident of their quality.

3.2   Simulations

This research leverages a dataset compiled and made publicly available in a GitHub repository from a prior study [21], which investigates the topological determinants of distributed averaging convergence rate in different graph families. The dataset encompasses 11828 sparse, fully-connected, undirected networks, generated using four distinct models: 1639 Erdős-Rényi graphs [7], 1642 geometric random graphs [17], 4070 scale-free graphs with adjusted clustering probabilities varying from 0.0 to 0.8 [9], and 4477 small-world graphs with rewiring probabilities ranging from 0.0 to 0.5 [25], all with sizes ranging between 200 to 1000 nodes and average degree up to 60.

The dataset comprises summary statistics, including averages, minimums, maximums, standard deviations, and skewness, for various global and local graph metrics. The selected metrics quantify global and local properties of the graph and capture a variety of topological features contributing to the convergence rate of the gossip algorithm. Global metrics provide insights into the overall state of the entire network, while local metrics offer more granular details by focusing on the immediate neighbourhood of each node. Global metrics encompass centralities (namely degree, betweenness, closeness, and eigenvector), eccentricity, and global efficiency, whereas local metrics include degree, clustering, and local efficiency. Additionally, the dataset includes other distance metrics (average shortest path and Wiener index), degree metrics (degree assortativity correlation and entropy degree), and spectral features represented by the two largest and two smallest eigenvalues of the adjacency and Laplacian matrices, amounting to a total of 48 graph metrics.

The dataset also reports the convergence rate of the distributed averaging gossip algorithm, estimated through event-driven simulations and averaged over 100 runs. The adopted gossip scheme realises a random neighbour selection and activates each node at the times of a rate 1 Poisson process so that each node initiates on average one interaction per unit of time.

3.3 Local Graph Averages

The current study focuses on the predictive capability of local metrics, as they can be computed by each node in its neighbourhood and do not require nodes to have any knowledge of the global structure of the graph, a realistic assumption in real networks. The focus is also on averages as summary statistics of local metrics, as they can be seamlessly computed in systems designed for distributed averaging without introducing further modifications to the communication protocol. The local metrics considered are the degree, the clustering coefficient and the local efficiency. The degree of a node v_i, here denoted by $k(v_i)$, is the cardinality of the neighbour set Ω_i and quantifies its connections within the network. The clustering coefficient $cl(v_i)$ is defined as the number of triangles passing through the node $T(v_i)$ divided by the number of possible triangles

$$cl(v_i) = \frac{2T(v_i)}{k(v_i)(k(v_i) - 1)},$$

and is a measure of the degree to which nodes tend to cluster together. The local efficiency $eff(v_i)$ is the average efficiency of all node pairs in the subgraph induced by the neighbours of v_i, where the efficiency of a node pair is the multiplicative inverse of the shortest path distance. It quantifies the resistance to failure on a small scale as it measures how effectively information is exchanged after removing the node [11]. The computation complexity of local metrics largely depends on network density. For very sparse graphs, local metrics can be computed by each node in constant time and a fully parallel fashion.

3.4 Regression Models

An Ordinary Least Squares (OLS) regression was employed to forecast the convergence rate of a gossip averaging algorithm using the averages of the three local metrics. The model is represented as:

$$\gamma = a \langle k \rangle + b \langle cl \rangle + c \langle eff \rangle + d, \tag{20}$$

where $\langle \rangle$ indicates the average of the corresponding local metric. The parameters were determined to be $a = 0.0016$ [0.0015, 0.0016], $b = -1.1069$ [-1.1163, -1.0973], $c = 0.3587$ [0.3499, 0.3676], and $d = 0.4066$ [0.4050, 0.4085]. The model achieved an R-squared value of 0.943, confirming that local metrics can effectively predict the convergence rate of gossip averaging algorithms across different graph topologies. The negative coefficient b suggests that highly clustered areas of the graph are less effective in propagating estimates because nodes are more likely to have shared neighbours passing redundant information. The positive coefficients a and c confirm the intuitive notions that more connections and more efficient information flows promote convergence.

The model's performance was benchmarked against other models built on different subsets of graph metrics potentially accessible to nodes under various conditions, as detailed in Table 1. These models were evaluated using performance metrics such as Mean Absolute Error (MAE), Mean Squared Error

(MSE), Root Mean Squared Error (RMSE), and R-squared (R^2). Under minimal assumptions, nodes can compute their degree, local efficiency, and clustering coefficient, and average these metrics through distributed averaging. They could then use the regression model (1) to estimate performance. In systems where nodes can also distributively compute maximum and minimum values, model (2) is applicable. If nodes are aware of the network's size, a relatively strong assumption in some networks, they can apply model (3), or model (4) if they also calculate distributed maximum and minimum values. In rare cases where nodes have full knowledge of the graph's structure, model (5) incorporating both local and global metrics can be utilised. This model, as detailed in previous work [21], predicts performance with remarkable accuracy. Analysis of models built on various combinations of graph metrics reveals that the most valuable global information for nodes is the network size.

Table 1. Performance metrics for regression models predicting the convergence rate of the averaging gossip algorithm from averages (avg), minimum (min), and maximum (max) values of local and global graph metrics.

Regression model	MAE	MSE	RMSE	R^2
(1) avg of local metrics	0.0317	0.0017	0.0408	0.943
(2) avg, min, max of local metrics	0.0278	0.0014	0.0368	0.954
(3) avg of local metrics + size	0.0240	0.0010	0.0316	0.966
(4) avg, min, max of local metrics + size	0.0171	0.0006	0.0241	0.980
(5) local and global metrics	0.0122	0.0003	0.0165	0.991

3.5   Distributed Accuracy Prediction

After validating the effectiveness of local metric averages in predicting the convergence rate of the averaging gossip protocol, a simulated distributed system was deployed. This system performed distributed averaging using the same gossip protocol detailed in Sect. 2.3 and leveraged in previous work to estimate the convergence rate in graphs with varying topologies [21]. The simulated distributed system was adapted for the intended tasks as follows:

1. Each node was assigned an accuracy value R_i. Throughout the experiments, nodes either shared the same value of R or were divided into two groups, each with different values of R, simulating systems where nodes fulfil different roles with distinct accuracy requirements.
2. Nodes were programmed to exchange information about their local neighbourhood with their neighbours, enabling each node to map their neighbours' neighbours and compute their local metrics (degree, clustering coefficient, and local efficiency).

3. Nodes performed distributed averaging on the three local graph metrics, either concurrently or before performing distributed averaging on their measured quantities.
4. Nodes leveraged the model defined in Eq. 20 to compute their current estimate of the convergence rate γ and applied the result, along with their set parameter R, in Eq. 19 to determine the time needed to achieve the desired level of accuracy t.
5. Nodes updated the estimate of t with each update of the local averages (if any) and compared it to the elapsed time to assess if the desired level of accuracy had been attained.

With respect to point 3, two principal strategies for estimating local metric averages were considered. The first strategy entails an initial round of distributed averaging to calculate the local metric averages, followed by a second round for propagating measured quantities. However, it requires a degree of coordination, possibly enabled by a distributed stopping protocol [18], where all nodes are synchronised to first perform distributed averaging on graph metrics for a set number of transmission cycles before applying distributed averaging on the measured quantities.

The second strategy involves performing distributed averaging of local metrics and measured quantities simultaneously. This method eliminates the need for the aforementioned coordination. The trade-off here is that the accuracy of the convergence rate prediction is directly tied to the accuracy of the estimated local metric averages. Over time, as nodes continually update their predictions based on increasingly precise local metric averages, the accuracy of these predictions generally improves, as they normally distribute around a value close to the actual time, with variance decreasing over time.

Empirical evaluation has shown that this method yields timely and accurate predictions, as they tend to stabilise at a time point t well before that time elapses, even in the least performing graphs. For instance, Fig. 1(a) depicts the distribution of node predictions of the time required to achieve an accuracy level of $R = 3$ in a geometric random graph of size $n = 1000$ and radius $r = 0.07$ made at different time points. It is shown that at time $t = 300$, more than 95% of predictions are found in the 650 ± 50 interval, so most predictions fall in a narrow interval around $t = 650$, the average time at which the desired level of accuracy was actually achieved in 100 runs of the averaging protocol.

Based on these observations, in the case studies that follow, the second implementation strategy was applied, propagating local metrics together with the measured values.

4   Applications

4.1   Topology Changes

Distributed averaging algorithms have been widely investigated in static topologies, characterised by fixed and reliable communication links throughout the

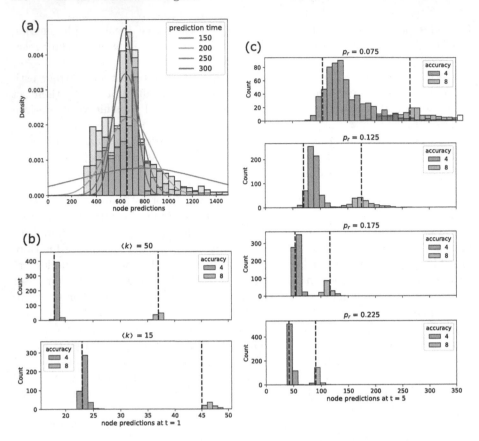

Fig. 1. (a) Distribution of node predictions of the time needed to achieve accuracy $R = 3$ made at different time points, and the corresponding best-fit normal distributions. (b) Distribution of node predictions in Erdős-Rényi graphs having fixed size $n = 500$ and decreasing average degree due to link failure. (c) Distribution of node predictions in small world graphs having fixed size $n = 800$, average degree $\langle k \rangle = 16$, and increasing rewiring probability p_r. In (b) and (c), 80% of nodes have R set to 4, while the remaining 20% have R equal to 8, modelling a network of agents with different accuracy requirements. In all graphs, the vertical lines represent the average times at which the desired level of accuracy was achieved in 100 runs of the gossip protocol.

observed time. This assumption is generally unrealistic for real networks, where the topology can differ in subsequent executions of the algorithm due to communication interference or agents changing positions [16]. However, agents are generally notified solely of changes in their immediate neighbourhood and are unable to assess topology changes on a large scale. The propagation and averaging of local metrics constitute a valuable tool for nodes to evaluate the convergence rate in the current communication network, especially in the event of rewiring and link failure, which can greatly affect performance.

Link Failure. Erdős-Rényi graphs are generally adopted to model scenarios where edges fail with equal probability f [8]. If the failure probability varies over time due to changes in the network communication medium, the graph topology is defined by a series of Erdős-Rényi graphs, each with edge probability $1 - f$. The method proposed in this study equips nodes with a tool to compute the convergence rate and, thus, assess the communication efficiency at any given time. Figure 1 (b) shows the distribution of node prediction in Erdős-Rényi graphs having fixed size $n = 500$ and decreasing average degree $\langle k \rangle$ due to link failure and demonstrates how predictions capture the lower convergence rate of a network that has lost over 70% of its links.

Rewiring. The convergence rate of averaging protocols can be dramatically increased without adding new links or nodes by means of *random rewiring* [25], leading to the design of small-world networks for ultra-fast consensus [13]. The proposed approach enables nodes to estimate the convergence rate of a network undergoing random rewiring at any given time. Figure 1 (c) presents the evolution of node predictions computed in small world graphs, generated according to the Watts-Strogatz model, with size $n = 800$, average degree $\langle k \rangle = 16$, and a progressively increasing rewiring probability p_r, demonstrating the adaptability of the proposed approach in dynamically reconfigured networks.

4.2   Anomaly Detection

Distributed averaging in networked systems allows individual nodes to evaluate how their sensing environment differs from that of the other nodes by comparing their measured value with their estimate of the global average. An anomalous value or outlier for the population can be defined as any measured quantity that is distant from the global average of a given amount M or a certain number of standard deviations m. The corresponding node can then raise an alarm to inform a remote control centre of the unexpected measurement, perform a corrective action, limit its interactions, or signal the neighbouring nodes of the lower reliability of its value.

Alarm System. In networks of sensing devices, any measurement distant at least M from the global average α can be considered anomalous by the system and trigger an alarm. Each node can deploy the proposed approach to evaluate the time t needed to attain its desired accuracy R_i and only then compare its measured value $x_i(0)$ with its estimated average $x_i(t)$. If $|x_i(0) - x_i(t)| > M$, the measurement is labelled as anomalous, and the node activates a response. This procedure is subject to *false negatives*, which do not detect anomalous measurements, and *false positives*, where regular values are erroneously detected as anomalous, because of the differences between the actual global average α and its local estimate at time t, $x_i(t)$. Notably, nodes face a trade-off between timely detection and classification error, meaning that a lower value of R_i enables faster but less accurate feedback.

Figure 2 (a) illustrates this concept in an Erdős-Rényi network with a size of $n = 1000$ and an average degree of $\langle k \rangle = 10$. In this network, the convergence rate of the averaging gossip algorithm is approximately $\gamma \approx 1/5$, implying that collective disagreement diminishes by a factor of 10 every 5 time units. The figure shows how classification errors of anomalous measurements decrease as the accuracy requirement R increases for two different distributions of initial values $x_i(0)$.

Outlier Identification. In sensing systems where measurements are expected to distribute normally with standard deviation σ, an outlier can be defined as any value distant more than m standard deviations from the group average α. From Eq. 16, 18 and 19, it follows that

$$-\log(\frac{\sigma(t)}{\sigma(0)}) = t\gamma \tag{21}$$

so that, at any time t, the standard derivation of the estimates $\sigma(t)$ can be derived from the initial deviation $\sigma(0)$ and the convergence rate of the graph γ, which can be estimated using the proposed methods. Each node v_i can then keep track of a confidence interval where the group average α is likely to be found. Since estimates are normally distributed, v_i has a high probability of finding the group average within three standard deviations from its estimate, i.e. $\mathbb{P}(\alpha \in ci_i) \approx 0.997$, where $ci_i = [x_i(t) - 3\sigma(t), x_i(t) + 3\sigma(t)]$. If the node initial value $x_i(0)$ is distant at least m standard deviations from this confidence interval, formally

$$\min_{\forall c \in ci_i} |x_i(0) - c| > m\,\sigma(0),$$

the measurement is detected as an outlier for the group, and the corresponding node initiates a response.

Figure 2(b) illustrates this principle in a scale-free communication network (with $n = 200, \langle k \rangle = 10, cl = 0.01$) where nodes are assigned initial values $x_i(0)$ from a normal distribution ($\mu = 0, \sigma = 1$). With a convergence rate $\gamma \approx 1/6$, the standard deviation of the estimates is expected to reduce by a factor of 10 every 6 time units. The results in Fig. 2(b)(i) illustrate how the standard deviation of the estimates evolves in accordance with Eq. 21, enabling nodes to ascertain if their initial value is an outlier within the population. In this scenario, all outliers are correctly identified by $t = 12$.

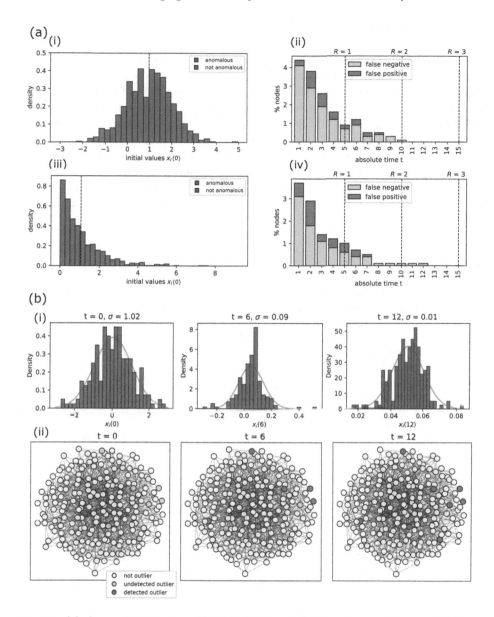

Fig. 2. (a) Anomaly detection. (i) Distribution of initial values $x_i(0)$ sampled from a normal distribution ($\mu = 1$, $\sigma = 1$), considered anomalous if $|\alpha - x_i(0)| > M$, with $M = 2$, and (ii) percentage of nodes making a classification error at time t. (iii) Distribution of initial values $x_i(0)$ sampled from a gamma distribution (shape $k = 1$, scale $\theta = 1$) with $M = 2$, and (iv) corresponding classification errors over time. (b) Outlier identification. (i) Distribution of the estimates x_i of the average of measured quantities with standard deviation evolving according to Eq. 21 and (ii) visual representation of the outlier detection process at times $t = 0, 6, 12$, with all outliers detected by $t = 12$.

5   Conclusion

The study proposes a distributed approach to estimate the convergence rate of an averaging gossip scheme deployed on a network. The key idea is to approximate the graph convergence rate with a linear combination of averages of local metrics. Nodes can then be programmed to estimate these parameters by distributed averaging and to implement the regression model in order to compute the convergence rate and the time needed to achieve the desired level of accuracy. The approach enables nodes to make informed decisions on their use of measured and estimated data and gain awareness of the global structure of the network. Future efforts will be directed toward identifying models able to provide more accurate predictions without increasing memory requirements or communication costs.

References

1. Assran, M., Loizou, N., Ballas, N., Rabbat, M.: Stochastic gradient push for distributed deep learning. In: International Conference on Machine Learning, pp. 344–353. PMLR (2019)
2. Boyd, S., Ghosh, A., Prabhakar, B., Shah, D.: Gossip algorithms: design, analysis and applications. In: Proceedings IEEE 24th Annual Joint Conference of the IEEE Computer and Communications Societies, vol. 3, pp. 1653–1664. IEEE (2005)
3. Boyd, S., Ghosh, A., Prabhakar, B., Shah, D.: Randomized gossip algorithms. IEEE Trans. Inf. Theory $52(6)$, 2508–2530 (2006)
4. Brust, M.R., Rothkugel, S.: Small worlds: strong clustering in wireless networks. arXiv preprint arXiv:0706.1063 (2007)
5. DeGroot, M.H.: Reaching a consensus. J. Am. Stat. Assoc. $69(345)$, 118–121 (1974)
6. Denantes, P., Bénézit, F., Thiran, P., Vetterli, M.: Which distributed averaging algorithm should I choose for my sensor network? In: IEEE INFOCOM 2008-The 27th Conference on Computer Communications, pp. 986–994. IEEE (2008)
7. Erdős, P., Rényi, A., et al.: On the evolution of random graphs. Publ. Math. Inst. Hung. Acad. Sci. $5(1)$, 17–60 (1960)
8. Hatano, Y., Mesbahi, M.: Agreement over random networks. IEEE Trans. Autom. Control $50(11)$, 1867–1872 (2005)
9. Holme, P., Kim, B.J.: Growing scale-free networks with tunable clustering. Phys. Rev. E $65(2)$, 026107 (2002)
10. Kempe, D., Dobra, A., Gehrke, J.: Gossip-based computation of aggregate information. In: 44th Annual IEEE Symposium on Foundations of Computer Science, 2003, pp. 482–491. IEEE (2003)
11. Latora, V., Marchiori, M.: Efficient behavior of small-world networks. Phys. Rev. Lett. $87(19)$, 198701 (2001)
12. Loizou, N., Richtárik, P.: Revisiting randomized gossip algorithms: general framework, convergence rates and novel block and accelerated protocols. IEEE Trans. Inf. Theory $67(12)$, 8300–8324 (2021)
13. Olfati-Saber, R.: Ultrafast consensus in small-world networks. In: Proceedings of the 2005, American Control Conference, 2005, pp. 2371–2378. IEEE (2005)
14. Olfati-Saber, R., Fax, J.A., Murray, R.M.: Consensus and cooperation in networked multi-agent systems. Proc. IEEE $95(1)$, 215–233 (2007)

15. Oliva, G., Panzieri, S., Setola, R., Gasparri, A.: Gossip algorithm for multi-agent systems via random walk. Syst. Control Lett. **128**, 34–40 (2019)
16. Patterson, S., Bamieh, B., El Abbadi, A.: Convergence rates of distributed average consensus with stochastic link failures. IEEE Trans. Autom. Control **55**(4), 880–892 (2010)
17. Penrose, M.: Random geometric graphs, vol. 5, OUP Oxford (2003)
18. Prakash, M., Talukdar, S., Attree, S., Patel, S., Salapaka, M.V.: Distributed stopping criterion for ratio consensus. In: 2018 56th Annual Allerton Conference on Communication, Control, and Computing (Allerton), pp. 131–135. IEEE (2018)
19. Qiu, L., Zhang, J., Tian, X.: Ranking influential nodes in complex networks based on local and global structures. Appl. Intell. **51**, 4394–4407 (2021)
20. Shames, I., Charalambous, T., Hadjicostis, C.N., Johansson, M.: Distributed network size estimation and average degree estimation and control in networks isomorphic to directed graphs. In: 2012 50th Annual Allerton Conference on Communication, Control, and Computing (Allerton), pp. 1885–1892. IEEE (2012)
21. Sirocchi, C., Bogliolo, A.: Topological network features determine convergence rate of distributed average algorithms. Sci. Rep. **12**(1), 21831 (2022)
22. Sirocchi, C., Bogliolo, A.: Community-based gossip algorithm for distributed averaging. In: Patiño-Martínez, M., Paulo, J. (eds.) Distributed Applications and Interoperable Systems: 23rd IFIP WG 6.1 International Conference, DAIS 2023, Held as Part of the 18th International Federated Conference on Distributed Computing Techniques, DisCoTec 2023, Lisbon, Portugal, June 19-23, 2023, Proceedings, pp. 37–53. Springer Nature Switzerland, Cham (2023). https://doi.org/10.1007/978-3-031-35260-7_3
23. Sundaram, S., Hadjicostis, C.N.: Distributed function calculation and consensus using linear iterative strategies. IEEE J. Sel. Areas Commun. **26**(4), 650–660 (2008)
24. Trinh, M.H., Ahn, H.S.: Theory and applications of matrix-weighted consensus. arXiv preprint arXiv:1703.00129
25. Watts, D.J., Strogatz, S.H.: Collective dynamics of 'small-world'networks. Nature **393**(6684), 440–442 (1998)
26. Xiao, L., Boyd, S.: Fast linear iterations for distributed averaging. Syst. Control Lett. **53**(1), 65–78 (2004)

Towards Graph Clustering for Distributed Computing Environments

Przemysław Szufel[✉][iD]

SGH Warsaw School of Economics, Warsaw, Poland
pszufel@sgh.waw.pl

Abstract. Several algorithms and tools that operate on graphs can significantly benefit from distributed computing. For instance, consider a logistic transportation network represented as a temporal graph. Optimizing transportation routes and times is a well-known NP-hard problem. One typical approach is problem decomposition, which requires optimal partitioning of the network. In such problems, the goals include minimizing the number of cross-partition edges, balancing the sizes of partitions, and controlling the number of partitions to match the capabilities of the computing environment. In this paper, we propose a mathematical formulation of the graph clustering problem for distributed computing environments, along with a simple initial heuristic that can be used to obtain partitions.

Keywords: Graph clustering · Distributed computing · Parallel computing · graph partitioning

1 Introduction

With the increasing number of CPU cores in computers and the widespread adoption of cloud computing, parallelization has become the primary method for achieving performance in processing significant amounts of data. Many algorithms and tools that work with graphs can gain considerable advantages from distributed computing. For example, consider a logistic transportation network modeled as a temporal graph. The optimization of transportation routes and schedules is a recognized NP-hard problem [9]. A common strategy for tackling this issue is to decompose the problem [7], which necessitates an optimal division of the network. In such a decomposition, the crucial factor is to minimize the level of interaction between the graph's partitions. Another example is node embedding algorithms such as Node2vec [4], which is based on random walks. These walks can mostly be computed in parallel on separated partitions in a distributed computing environment. The crucial thing to minimize is again the

This research is supported by the Polish National Agency for Academic Exchange under the Strategic Partnerships programme, grant number BPI/PST/2021/1/ 00069/U/00001. This research was inspired by visits to Loyola University Chicago during the grant programme.

© The Author(s), under exclusive license to Springer Nature Switzerland AG 2024
M. Dewar et al. (Eds.): WAW 2024, LNCS 14671, pp. 146–158, 2024.
https://doi.org/10.1007/978-3-031-59205-8_10

number of cross-partition links, as these cross-partition walks are the main factor that limits the parallelization of the algorithm.

The advancement of multi-processing libraries, technologies, and programming languages that natively support distributed computing, such as Julia [2], presents an opportunity for developing scalable, parallel implementations of new algorithms for graphs. This endeavor becomes particularly challenging when combined with multiprocessing and distributed computing. Such approaches may be necessary in scenarios where the graph is too large to fit into the memory of a single machine due to a vast amount of metadata, or when the performed analysis requires significant computational resources. In these instances, the graph is partitioned into smaller subgraphs, and computations are performed on each partition separately.

In the process of distributed computing, the major processing costs are incurred when cross-process communication occurs. In the case of graphs partitioned in a computer's memory, this means communication between partitions. Consider the graph presented in Fig. 1, which has been divided into two partitions, each being processed by a separate (perhaps remote) process. For several algorithms with a graph-based representation of data (such as the aforementioned route planning or node2vec), most of the processing could happen mostly on the independent partitions, except for the cross-partition link. Hence, the main challenge for this type of parallel graph processing scenarios is to identify a partitioning that minimizes the number of cross-partition links. This is crucial because the cross-partition links require communication between the processes performing the computations. Since inter-process communication is a costly operation, it is the main factor that limits the scalability of the computations. The second goal is to balance the workload received by each process. The results presented in [12] note that in parallelizable computing, the overall completion time is determined by the longest workload. They discuss an example where processing time variance can have massive consequences for the overall processing time. Hence, the second crucial property is balancing the size of graph partitions. In addition to controlling partition size and minimizing the number of cross-partition links, a third important factor is to limit the total number of partitions to align with the computational resources available. The number of partitions should be controlled to match the capabilities of the computing environment.

The measurement of subgraph size can vary significantly depending on the specific application—the size might include just the number of vertices, or it could also encompass the number of edges or the size of the metadata associated with a node or edge. The aim of this work is to propose a mathematical formulation of the graph clustering problem for distributed computing environments that captures these crucial performance aspects. We also present the mixed integer linear programming (MILP) approach, which is common to other graph clustering methods as seen in [10] or [1]. The integer programming representation can be solved with mathematical programming solvers such as Gurobi [5]. However, solvers can find solutions only for small scale problems and hence we propose a simple initial heuristic that can be used to obtain partitions. We have developed a prototype of the proposed algorithm and evaluated it on a collection of artificially generated graphs as well as on a real-world dataset representing a transportation network.

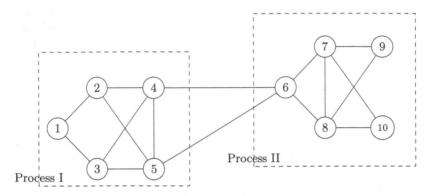

Fig. 1. Example of a graph divided into two partitions. With a minimal number of links between these partitions, computations within each partition can proceed in parallel, each handled by a distinct process.

2   Model

Let $G = (V, E)$ be an undirected graph with a set of nodes V and a set of vertices E. Let u_v be a weight of vertex $v \in V$. Let R be a set of partitions of G and s_{vr} be a binary variable indicating whether vertex $v \in V$ belongs to partition $r \in R$. Let us define the subset of nodes belonging to partition r as $V_r = \{v \in V : s_{vr} = 1\}$, $V_r \subset V$. We assume that each vertex can belong to a single partition at any time that is:

$$\sum_{r \in R} s_{vr} = 1, \qquad \forall v \in V.$$

Let $|V|$ denote the total number of vertices and $|V_r|$ the number of vertices in the partition $r \in R$. Let a_{vw} be a binary variable indicating whether there is an edge between vertices v and w. Let us define the subset of edges between partitions r and r' as $E_{rr'} = \{(v, w) \in E : s_{vr} = 1, s_{wr'} = 1\}$, $E_{rr'} \subset E$. Let $c_{rr'}$ be a weight of the edge between partitions r and r'. Let z be the size of the biggest partition in a given partitioning with respect to vertex weights u_v:

$$z = max \left\{ \sum_{v \in V} u_v s_{vr}; \quad r \in R \right\}. \tag{1}$$

If $u_v = 1$ this can be just written as $z = max\{|V_r|; r \in R\}$.

Let K^* be the target desired number of partitions. Let u^* be the target average partition size with respect to vertex weights u_v

$$u^* = \frac{1}{K^*} \sum_{v \in V} u_v. \tag{2}$$

Let $\alpha, \beta, \gamma, \delta \geq 0$ be weights that scale the components of the objective function. The objective function is defined as follows:

$$\text{Min} \quad \alpha \sum_{v,w \in V} \left(a_{vw} \left(1 - \sum_{r \in R} s_{vr} s_{wr} \right) \right) + \beta * z +$$

$$\gamma \sum_{r \in R} \left(u^* - \sum_{v \in V} s_{vr} \right)^2 - \delta \sum_{v,w \in V} \left(a_{vw} \sum_{r \in R} s_{vr} s_{wr} \right)$$

The first component (minimized) of the objective function is the number of edges between partitions, weighted by α. The second (minimized) component is the size of the largest partition, weighted by β. The third component (minimized) is the variance of partition sizes, weighted by γ. The fourth component (maximized) is the number of edges within partitions, weighted by δ. These four components constitute the objective function.

Given the prioritization of minimizing the number of cross-partition links and keeping the size of the largest partition small, we simplify the objective function by setting $\gamma = 0$ and $\delta = 0$. This simplification leads us to focus on an objective function that only minimizes the number of cross-partition links and the size of the largest partition.

$$\text{Min:} \alpha \sum_{v,w \in V} \left(a_{vw} \left(1 - \sum_{r \in R} s_{vr} s_{wr} \right) \right) + \beta * z \tag{3}$$

The objective function described above, in Eq. 3, is minimized subject to the following constraints:

$$\sum_{v \in V} s_{vr} u_v \leq z \qquad\qquad \forall\, r \in R \tag{4}$$

$$z \leq \zeta u^* \tag{5}$$

$$\sum_{r \in R} s_{vr} = 1 \qquad\qquad \forall\, v \in V \tag{6}$$

$$s_{vr} \in \{0,1\} \qquad\qquad \forall\, v \in V, r \in R \tag{7}$$

$$z \geq 0 \tag{8}$$

The first constraint ensures that the weighted size (using node weights u_v) of each partition is smaller than z. The second constraint requires that the size of the largest partition does not exceed the average partition size u^*, scaled by a factor of ζ. The third constraint guarantees that each vertex is assigned to exactly one partition. The fourth constraint specifies that the values of s_{vr} must be binary, indicating whether a vertex v belongs to partition r. The last constraint mandates that the value of z must be non-negative. It is important to note that the maximum number of partitions is denoted by $|R|$, reflecting real-world situations where the number of partitions is constrained by the availability of computing resources. Nevertheless, if the parameter ζ is sufficiently large, the actual number of partitions in the optimal partitioning solution may be less than $|R|$.

The proposed optimization model is formulated as a mixed integer linear programming (MILP) problem. It is important to note that the objective function, as presented in Eq. 3, is linear, and all the constraints introduced are also linear. This characteristic allows the problem to be solvable using mathematical programming solvers. We have implemented the proposed MILP model using the Julia programming language, in conjunction with the JuMP.jl library for mathematical optimization, and the Gurobi optimization library [5]. A sample solution, utilizing the well-known Zachary's karate club graph, is illustrated in Fig. 2.

Fig. 2. Sample partitioning of the Zachary's karate club graph into $|R| = 4$ clusters for $\alpha = 1$, $\beta = 0$ and $\zeta = 1.2$.

Unfortunately, it has been found that the mathematical programming approach is only feasible for small graphs. For larger graphs, the number of variables and constraints becomes too large for the solver to handle effectively. The exponential growth in computational time is clearly depicted in Fig. 3. This challenge arises because the optimization model includes several binary variables, which are known to significantly increase the difficulty of finding a solution in practice. To address the issue of computational complexity, we propose a simple heuristic that can be used to obtain partitions (Sect. 3).

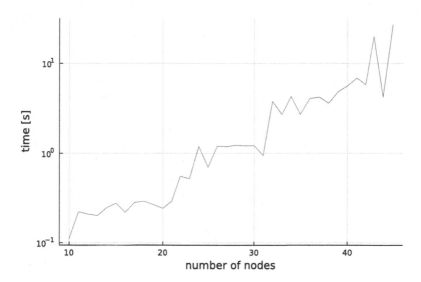

Fig. 3. Computational time of the Gurobi solver for various sizes of the Watts-Strogatz graph with vertices of degree 6, a rewiring probability of 0.05, and the number of partitions $|R| = 6$. The computational time increases exponentially with the number of vertices.

3 Heuristic

In this section, we propose a heuristic that can be used to obtain partitions, inspired by the Louvain algorithm [3]. The heuristic is based on the idea of starting with each vertex in its own partition. We then iteratively merge partitions to minimize the number of cross-partition links until the number of partitions equals the target number of partitions K^*. The merges are performed in such a way that the size of the resulting partition does not exceed the maximum partition size, defined as ζu^*.

It's important to note that in the actual implementation, the matrix s is not stored in memory. Instead, the indices of non-zero size clusters are stored in a vector, as each node must be assigned to exactly one partition at any given time. For clarity, we are representing s as a matrix. Additionally, it's worth noting that data science-oriented programming languages, such as Julia, allow for the exposition of such data structures as a matrix of one-hot rows by implementing appropriate interfaces (in Julia, this would be the `AbstractMatrix{Bool}` data type).

The heuristic is outlined in Algorithm 1. The algorithm involves the following steps: First, we initialize the node index matrix **s** with each node in a separate partition (Line 2). Second, we randomly select a cluster $r_1 \in R$ that is non-empty and below the size limit ζu^* (Line 8). Next, we identify all clusters that share at least one edge with cluster r_1 (Line 12). This step could also include Next, we identify the best partition r^* to merge into r_1 (Line 15). Note that in the process of identifying the best partition we simultaneously take into consideration the reduction of the cross-partition edges with weight α as well we try to avoid increasing the largest cluster's size with the weight β. If r^* is not empty we merge partitions r_1 and r^* (Line 18). After the merging the vertex indices V_{r_1} and V_{r_2} need to be updated (Line 22). We repeat these steps until the number of partitions matches the target number of partitions K^* or the maximum number of steps t^* is reached. Additionally, the algorithm can be terminated when the number of cross-partition links equals zero. The resulting solution **s** will contain empty partitions, and the zero-columns are removed (Line 26).

The proposed algorithm is inspired by the Louvain method [3], however since the goal is only to minimize the number of cross-partition links we use a different problem formulation. Rather than consider aggregated vertices with self-loops and multi-edges we consider the original graph. This results in a different representation of clusters without the need for vertex aggregation. The algorithm is also different that the maximum vertex-weighted size of the merged partition is limited to ζu^*.

Note that the proposed algorithm is not guaranteed to converge to the optimal solution. Hence, a solution obtained with the proposed heuristic can be further refined by solving the MILP model presented in the previous section. The solution from the heuristic can be set as the starting point for the MILP solver.

We have developed a prototype of the proposed heuristic in the Julia programming language. In the following section, we present the results from the numerical experiments conducted with this prototype.

4   Experiments

In this section, we perform numerical testing of Algorithm 1. The heuristic has been implemented in the Julia programming language. We benchmark the heuristic against implementations of the Louvain [3] and Leiden algorithms [11], which are designed to maximize the modularity of a graph. It's important to note the significant difference between modularity and the objective function proposed in this paper, as delineated in Eq. 3. Modularity measures the quality of a graph's partitioning, considering both the number of inter-cluster connections and the density of within-cluster connections. However, it does not focus on balancing cluster sizes or controlling the total number of clusters, though modifications to these algorithms can introduce such considerations by adding a stop condition. The objective function in Eq. 3 permits a partition to include two disconnected communities without imposing a penalty, which contrasts with modularity-based

Algorithm 1. Graph clustering heuristic minimizing the intercluster connections

---

Require: $G = (V, E)$: Graph
Require: K^*: Target number of partitions
Require: α: Weight of the number of cross-partition links
Require: β: Weight of the largest partition size
Require: ζ: Maximum partition size scaling factor
Require: t^*: Maximum algorithm number of steps
Ensure: $\mathbf{s}$: Node index matrix of size $|V| \times |R|$
Ensure: $|R| = |V|$: initial number of partitions when each node is in a separate partition
Ensure: $K = |R|$: the current number of partitions

1: # Initialize node-to-partition matrix $\mathbf{s}$:
2: $s_{vv} = 1 \quad \forall_{v \in V}$
3: Initialize iteration step counter $t = 1$
4: **while** $K > K^* \wedge t < t^*$ **do**
5: $\quad$ Define nonempty partitions $R^* = \{r \in R : |V_r| \geq 1\}$
6: $\quad$ # Set the smallest and largest non-empty partition sizes:
7: $\quad$ $u^{(min)} = \min_{r \in R^*} |V_r|$ and $u^{(max)} = \max_{r \in R^*} |V_r|$
8: $\quad$ Select a random smallest sized partition $r_1 \in R^*$ such as $|V_{r_1}| = u^{(min)}$
9: $\quad$ # Find all partitions that share at least one edge with the partition r_1
10: $\quad$ and there is space to combine partitions: $\mathbf{R}^{(r_1)} =$
11: $\quad\quad \{r_2 \in R^* : r_2 \neq r_1 \wedge \left(\sum_{v \in V_{r_2}} u_v \right) \leq \zeta u^* - \left(\sum_{v \in V_{r_1}} u_v \right) \wedge$
12: $\quad\quad \wedge \exists_{v_1 \in R_{r_1}, v_2 \in R_{r_2}} (v_1, v_2) \in E\}$
13: $\quad$ # Identify the best partition r^* to merge into r_1:
14: $\quad$ $r^* = \arg\max_{r_2 \in \mathbf{R}^{(r_1)}} \alpha |\{e = (v_1, v2) : v_1 \in V_{r_1} \wedge v_2 \in V_{r_2}\}| +$
15: $\quad\quad\quad\quad\quad\quad -\beta \max(0, |V_{r_1}| + |V_{r_2}| - u^{(max)})$
16: $\quad$ **if** $r_2 \neq \emptyset$ **then**
17: $\quad\quad$ # Merge partitions r_1 and r^*:
18: $\quad\quad$ **for** $v \in V_{r_1}$ **do**
19: $\quad\quad\quad$ $s_{vr_1} = 0$
20: $\quad\quad\quad$ $s_{vr^*} = 1$
21: $\quad\quad$ **end for**
22: $\quad\quad$ Update sets V_{r_1} and V_{r^*}
23: $\quad$ **end if**
24: $\quad$ $t = t + 1$
25: **end while**
26: Remove empty partitions from $\mathbf{s}$
27: Post-optimize via a MILP solver by utilizing $\mathbf{s}$ as the initial solution.
28: **return** $\mathbf{s}$

---

methods. This approach meets the requirements of parallel computing. Given their popularity, the Louvain and Leiden algorithms were selected for benchmarking. We utilized the CDlib library in Python for running the Louvain and Leiden algorithms [8]. Due to the non-deterministic nature of the Louvain and Leiden algorithms as well as the heuristic presented in Algorithm 1, we conducted 100 runs for each and reported the average results. The final step of the heuristic involves post-optimization through the injection of the solution s into a MILP solver within a specified time limit. Although the problem is NP-hard, the MILP solver can identify minor adjustments to the solution s that may slightly enhance the solution quality.

4.1 Performance on the Karate Graph

First, we begin with Zachary's Karate Club graph, with a sample partitioning illustrated in Fig. 2. The findings are summarized in Table 1. It is important to highlight that for the target number of partitions $|R|$, we selected $|R| = 4$, a value that the Louvain and Leiden algorithms almost invariably produce for this graph. The results indicate that the proposed heuristic is capable of achieving partitions that serve as a solid foundation for subsequent processing. Notably, the variance in partition size resulting from Algorithm 1 is significantly reduced, and the average partition size may also be smaller. Evidently, the MILP approach, as an exact method, delivers the most satisfactory solution.

Table 1. Results for Zachary's Karate Club graph indicate that the proposed heuristic, for positive β values, enables control over the size of the partitions. It is particularly noteworthy that the variance in partition size achieved through Algorithm 1 is significantly less than the variance in partition size produced by the Louvain and Leiden algorithms. Additionally, the last column presents the number of cross-partition links.

Method	β	Average $|R|$	Std. Dev. of part size	Max part size	Cross-partition
Heuristic	0.0	4.35	4.14	13.37	29.41
Heuristic	5.0	4.01	2.608	11.6	33.06
Heuristic	10.0	4.01	2.636	11.54	33.41
Louvain	—	4.0	3.839	12.49	20.89
Leiden	—	4.0	3.512	12.0	21.0
MILP/Gurobi	0.0	4.0	17.0	34.0	0.0
MILP/Gurobi	5.0	4.0	2.38	10.0	24.0
MILP/Gurobi	10.0	4.0	1.0	9.0	27.0

4.2 Performance on the ABCD Graph

The ABCD algorithm, as proposed in [6], facilitates the generation of graphs with specified community numbers and sizes, and allows control over the density of cross-partition links. We generated 100 graphs with $|V| = 50$ vertices and an average of $|E| = 93$ edges, and another 100 graphs with $|V| = 100$ vertices and an average of $|E| = 185$ edges. The mixing parameter used was $\xi = 0.25$, and the number of communities was set to $|R| = 5$. The findings are summarized in Table 2. These results indicate that the heuristic can achieve a lower variance in partition sizes compared to the Louvain and Leiden algorithms. However, the heuristic appears to be less efficient for graphs at their maximum size, particularly for larger graphs.

Table 2. Results for a set of randomly generated ABCD graph. For each of 100 graphs of given sizes the algorithms have been run once. Note that the last column contains the number of cross-partition links.

Method	β	$|V|$	Average $|R|$	Std. Dev. of part size	Max part size	Cross-partition
Heuristic	0.0	50	5.0	5.299	18.01	30.87
Heuristic	5.0	50	5.0	2.45	13.34	37.27
Heuristic	10.0	50	5.0	2.451	13.35	37.32
Louvain	—	50	5.93	2.779	12.38	31.01
Leiden	—	50	5.91	2.495	11.97	29.61
Heuristic	0.0	100	5.0	13.31	41.3	55.4
Heuristic	5.0	100	5.0	4.482	26.28	71.77
Heuristic	10.0	100	5.0	4.459	26.23	72.1
Louvain	—	100	8.31	4.033	18.34	64.01
Leiden	—	100	7.98	3.733	18.54	59.97

4.3 Performance on a Road Network

In this experiment, we focus on a road network in central Boston. The data was obtained from the Open Street Map project using the Overpass API. We processed the data into a connected directed graph utilizing the OpenStreetMapX.jl library and then transformed the directed graph into an undirected one. We consider one smaller and one larger network of central Boston with the final graph consisting of (a): $|V| = 334$ vertices and $|E| = 412$ edges and $|V| = 3621$ vertices and $|E| = 7220$ edges respectively — see Fig. 4.

The results are summarized in Table 3. Given that the Louvain and Leiden algorithms typically yield approximately 15 partitions for the small map and approximately 40 partitions for the large Boston map, we have adjusted the proposed heuristic to generate a comparable number of partitions to ensure comparability. On this graph, the heuristic underperforms relative to the Leiden algorithm, likely due to its lack of a mechanism to prevent the relocation of critical nodes between partitions. In future research, we aim to remedy this issue by implementing a post-optimization mechanism.

156   P. Szufel

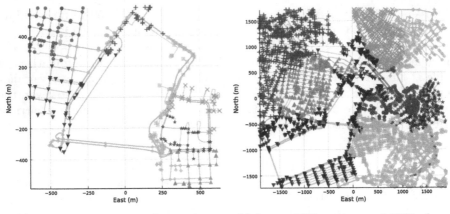

(a) Small: 334 vertices and 598 edges   (b) Large: 3621 vertices and 7220 edges

Fig. 4. A graph representing a road network in central Boston, used for evaluating the clustering heuristic described in Algorithm 1. Here the Algorithm 1 was used to cluster the small and large graphs into $K^* = 9$ partitions (vertices are corresponding to intersections).

Table 3. Results for the Boston map indicate that for small maps the Leiden algorithm performs significantly better compared to the heuristic. However, for larger graph the proposed heuristic leads to smaller size of the biggest partition and significantly slower standard deviation of partition sizes.

Map	Method	β	Average $|R|$	Std. Dev. of part size	Max part size	Cross-part
Small	Heuristic	0.0	15.0	9.862	45.23	44.7
Small	Heuristic	5.0	15.0	6.022	33.62	51.55
Small	Heuristic	10.0	15.0	6.022	33.62	51.55
Small	Louvain	—	15.34	7.753	36.74	38.66
Small	Leiden	—	15.74	5.838	34.0	38.22
Large	Heuristic	0.0	40.0	39.65	220.0	239.7
Large	Heuristic	5.0	40.0	21.76	140.1	301.0
Large	Heuristic	10.0	40.0	21.76	140.1	301.0
Large	Leiden	missing	40.0	34.89	160.0	176.0
Large	Louvain	missing	39.9	36.07	163.0	197.2

The numerical experiments demonstrate that the heuristic can slightly outperform the Louvain algorithm in terms of balancing cluster sizes and performs comparably to the Leiden algorithm for the smaller graph. This discrepancy arises because the heuristic lacks a mechanism, present in the Leiden algorithm, for preventing vertex movements that could affect the community structure. However, for larger graph the proposed heuristic leads to smaller size of the biggest partition and significantly slower standard deviation of partition sizes.

In practice this would lead to more evenly distributed computational workflows. It's important to note that the last column displays the number of cross-partition links. Moreover, it's important to note that the other algorithms do not allow for the specification of the number of clusters. Thus, our comparison has been based on a worst-case scenario—the number of partitions most frequently generated by Louvain and Leiden. Unlike these algorithms, the proposed heuristic accepts the number of partitions as a parameter. Additionally, it has been observed that Leiden significantly outperforms Louvain in minimizing cross-partition connections. Therefore, the next step involves incorporating some mechanisms from Leiden into the proposed heuristic, which will be explored in further research.

5   Conclusions

In this paper, we have proposes a mathematical formulation of the graph clustering problem for distributed computing environments, accompanied by a straightforward initial heuristic for obtaining partitions. We have implemented a prototype of the proposed algorithm and tested it on a set of artificially generated graphs as well as on a real-world dataset representing a transportation network.

The results demonstrate that the proposed heuristic can achieve partitions serving as a baseline for further processing. A potential enhancement for the developed heuristic involves extending the algorithm to address the issue identified by the authors of the Leiden algorithm [11], namely that the Louvain algorithm tends to shuffle vertices, detrimentally affecting the community structure. Since the current version of our algorithm builds upon the Louvain algorithm and merely aggregates partitions, it is similarly susceptible to this issue. However, our proposed heuristic offers configurability for any number of target partitions, a flexibility not afforded by either the Louvain or Leiden algorithms.

Future work will focus on refining the algorithm to ensure that nodes critical to a given partition are not relocated to another partition, which is expected to substantially enhance the model's properties.

References

1. Aref, S., Chheda, H., Mostajabdaveh, M.: The Bayan algorithm: Detecting communities in networks through exact and approximate optimization of modularity. arXiv preprint arXiv:2209.04562 (2022)
2. Bezanson, J., Karpinski, S., Shah, V.B., Edelman, A.: Julia: A fast dynamic language for technical computing. arXiv preprint arXiv:1209.5145 (2012)
3. Blondel, V.D., Guillaume, J.L., Lambiotte, R., Lefebvre, E.: Fast unfolding of communities in large networks. J. Stat. Mech: Theory Exp. **2008**(10), P10008 (2008)
4. Grover, A., Leskovec, J.: node2vec: scalable feature learning for networks. In: Proceedings of the 22nd ACM SIGKDD International Conference on Knowledge Discovery and Data Mining, pp. 855–864 (2016)
5. Gurobi Optimization, LLC: Gurobi Optimizer Reference Manual (2021), https:// www.gurobi.com

6. Kamiński, B., Prałat, P., Théberge, F.: Artificial benchmark for community detection (ABCD)-fast random graph model with community structure. Netw. Sci. **9**(2), 153–178 (2021)
7. Nowak, M., Szufel, P.: Technician routing and scheduling for the sharing economy. Eur. J. Oper. Res. **314**(1), 15–31 (2024)
8. Rossetti, G., Milli, L., Cazabet, R.: CDLIB: a python library to extract, compare and evaluate communities from complex networks. Appl. Netw. Sci. **4**(1), 1–26 (2019). https://doi.org/10.1007/s41109-019-0165-9
9. Safra, S., Schwartz, O.: On the complexity of approximating tsp with neighborhoods and related problems. Comput. Complexity **14**, 281–307 (2006)
10. Sato, K., Izunaga, Y.: An enhanced MILP-based branch-and-price approach to modularity density maximization on graphs. Comput. Oper. Res. **106**, 236–245 (2019)
11. Traag, V.A., Waltman, L., Van Eck, N.J.: From Louvain to Leiden: guaranteeing well-connected communities. Sci. Rep. **9**(1), 5233 (2019)
12. Wang, D., Joshi, G., Wornell, G.: Efficient task replication for fast response times in parallel computation. In: The 2014 ACM International Conference on Measurement and Modeling of Computer Systems, pp. 599–600 (2014)

HypergraphRepository: A Community-Driven and Interactive Hypernetwork Data Collection

Alessia Antelmi[1], Daniele De Vinco[2](✉), and Carmine Spagnuolo[2]

[1] Università degli Studi di Torino, Turin, Italy
alessia.antelmi@unito.it
[2] Università degli Studi di Salerno, Fisciano, Italy
{ddevinco,cspagnuolo}@unisa.it

Abstract. Hypergraph research has been thriving over the past few years, with a growing interest in a plethora of domains. Despite this remarkable surge, the lack of a comprehensive platform for searching and downloading diverse and well-curated datasets poses a significant obstacle to the continued advancement of the field. This absence hinders the ability of researchers and practitioners to validate and benchmark their hypergraph algorithms and models effectively.

To bridge this gap, we present HypergraphRepository, a web-based data collection aiming to serve as a centralized hub for hypergraph datasets, fostering collaboration and knowledge exchange within the hypergraph research community. In this paper, we detail the platform's architecture, features, and the collaborative framework it offers. HypergraphRepository is a GitHub-based open-source project.

Keywords: Hypergraph repository · Community-driven · Interactive · Web-based platform · Network collection

1 Introduction

Hypergraphs are the natural representation of a broad range of systems where high-order relationships exist among their interacting parts. In technical terms, a hypergraph extends the concept of a graph, allowing a (hyper)edge to connect an arbitrary number of nodes [11]. This extension proves invaluable when dealing with systems exhibiting highly non-linear interactions among their constituents [6]. In practice, hypergraphs become essential for modeling complex group interactions that cannot be adequately described through dyads (and, hence, via graphs). This modeling approach is particularly apt for capturing social systems where individuals engage in groups of varying sizes [23]. For instance, in the case of a co-authorship collaboration network, a hyperedge may represent an article and link together all authors (nodes) having collaborated on it [3]. Further, hypergraphs prove useful for embedding sociological concepts to

M. Dewar et al. (Eds.): WAW 2024, LNCS 14671, pp. 159–173, 2024.
https://doi.org/10.1007/978-3-031-59205-8_11

explore the dynamics of opinion formation [27] and social influence diffusion [28] when groups are explicitly taken into account, as well as for modeling epidemic-spreading processes to expressly consider community structure [8], and group dynamics [5]. Similar situations involving high-order interactions are evident in biology, ecology, and neuroscience [6].

In contrast to graphs, the literature on hypergraphs is in its adolescence thanks to a recent rise of systematic studies demonstrating how the transformation of a hypergraph to a classical graph either leads to an unavoidable loss of information or creates a large number of extra nodes/edges that increases space and time requirements in downstream graph analytic tasks [3]. This recent attention toward hypergraphs is reflected in the absence of one or more established repositories that offer readily available benchmark hypergraphs. The lack of standardized benchmark datasets may pose a challenge for researchers aiming to assess and compare claims, hypotheses, and algorithms specifically designed for hypergraphs in different application scenarios. Addressing this gap could potentially contribute to the growth of hypergraph research and foster a more robust understanding of their applications and implications across diverse fields while favoring the open science principles [12]. To this end, our paper introduces *HypergraphRepository*, the first open-source, community-driven, and interactive hypergraph collection. Our project stands on two fundamental pillars. First, its primary objective is to establish a dedicated space crafted by and for the community, allowing users to contribute by uploading their own datasets. This collaborative approach ensures that the repository evolves organically, reflecting the diverse needs and interests of the hypergraph research community. Second, the repository's interactive design aims to facilitate the exploration and comparison of a broad spectrum of datasets. The contribution of our paper is two-fold and can be summarized as follows:

- We introduce *HypergraphRepository*, the first systematic hypergraph data collection, where researchers can quickly and interactively compare, explore, and analyze data in real-time via a web-based platform. HypergraphRepository is open-source and available at hypergraphrepository.di.unisa.it.
- We provide an open-source community-driven data repository where users can contribute by sharing their own data and insight through a system of pull requests handled via git. The hypergraph database is available at github.com/HypergraphRepository/datasets.

The remainder of the paper is organized as follows. Section 2 reviews existing network collections. Section 3 describes HypergraphRepository, detailing the dataset creation and uploading pipeline and the currently available analytics features. Section 4 concludes this work, giving an overview of the current work and potential directions for further development.

2   Related Work

This section reviews existing network repositories.

Graph Repositories. The most famous graph repositories are probably SNAP [21] and Network Repository [26]. The `SNAP collection`, managed by the Stanford Graph Learning Research Group, encompasses more than 200 real-world network datasets from diverse domains and of various types. These datasets are provided in multiple formats, ensuring their compatibility with a broad spectrum of graph analysis tools. Launched in July 2009, the SNAP website primarily hosts datasets gathered to serve the specific research objectives of the research group. `ch1NetworkRepository` is a web-based data repository for interactively exploring, visualizing, and comparing a large number of networks. This platform also integrates social and collaborative features, enabling users to engage in discussions, share observations, and exchange visualizations related to each network. Currently, the repository contains more than 5000 networks sourced from 19 diverse categories (e.g., social, biological). These collections cover various network types, such as bipartite and time-series networks, and span domains like social sciences, physics, and bioinformatics. Both repositories come with their own programmatic libraries or tools for graph analysis and possible support for machine learning frameworks.

The `KONECT Project` [18,19] is another example of a platform where users can download network datasets and visualize their statistics online. The project is run by Jérôme Kunegis, and the entire source code is available as free software. The repository also includes a network analysis toolbox for GNU Octave, a network extraction library, as well as code to generate all statistics and plots shown on the website. Currently, The KONECT project boasts a collection of 1,326 network datasets in 24 categories.

Two websites where it is possible to download large graph datasets are the website of the `Laboratory for Web Algorithmics` [9,10] and Amazon's project `ch1GraphChallenge` [1]. The former offers a diverse array of graph categories accompanied by comprehensive statistics, including plots detailing degree distribution, the size of the giant component, average and median distance, as well as the harmonic diameter. The latter refers to the Graph Challenge data sets available to the community free of charge as part of the AWS Public Data Sets program.

The `Open Graph Benchmark` (OGB) [15] and the `Illinois Graph Benchmark` (IGB) [16] are two recent yet already mature projects tailored for advancing graph machine-learning tasks, aiming to facilitate the creation of novel models and the comparison with existing approaches. The OGB repository covers diverse scale graphs from various domains, such as biological networks, molecular graphs, academic networks, and knowledge graphs. This project fully automates dataset processing, providing graph objects that are fully compatible with Pytorch Geometric [13] and DGL [29], as well as standardized dataset splits and evaluators that allow for easy and reliable comparison of different models in a unified manner. The IGB project specifically focuses on training and evaluating graph neural network models. In particular, it provides highly labeled graphs, including both homogeneous and heterogeneous large-scale real-world citation graphs, with more than 40% of their nodes labeled. IGB is open-sourced and, like OGB, supports DGL and Pytorch Geometric frameworks.

Hypergraph Repositories. As of our current knowledge, there is a lack of a comparable repository dedicated to hypergraphs. The limited online resources typically consist of personal web pages where authors provide downloadable datasets used in their own work. Although this practice contributes to the reproducibility of their research, it falls short of establishing a universal platform. Moreover, this approach does not guarantee the expansion of dataset availability, as it relies on individual efforts. The most notable example is Professor Austin R. Benson's personal website [7]. Among other datasets, on this webpage, users can access 17 temporal hypergraphs, 11 node-labeled hypergraphs, 10 edge-labeled hypergraphs, and 2 hypergraphs labeled with a core-fringe structure. A short list of resources can also be found in the following pages [17, 22, 24, 25].

3   HypergraphRepository

HypergraphRepository stands as the first open-source, community-driven web platform designed to interactively explore, compare, and download hypergraph data. The ultimate objective of our platform is to assist users throughout the workflow of hypergraph-related tasks by serving as a centralized hub where they can access various hypergraph types (e.g., temporal, directed, weighted interactions) from diverse application domains (e.g., social, transportation, biological networks). This approach is designed to streamline the initial stages of hypergraph processing, providing users with readily available hypergraphs tailored for different tasks, such as community detection, hyperedge prediction, or node classification/ranking.

The dual nature of being both open-source and community-driven opens up a two-fold avenue for user contribution. First, users can actively participate in dataset creation, modification of existing datasets and take on roles as community reviewers. Second, users have the opportunity to contribute directly to the platform itself by engaging in discussions about potential updates to the repository. This collaborative process empowers users to request and add new features to the website, including the integration of novel interactive plots. The entire workflow, from dataset management to platform enhancements, operates seamlessly through the GitHub platform, fostering a transparent and collaborative environment. Specifically, users can utilize the pull request system on GitHub to propose modifications and improvements. This strategy not only promotes accessibility but also enhances the overall quality and integrity of the datasets. By implementing a stringent review mechanism as part of the submission process, the platform ensures that the available hypergraphs adhere to high standards driven by the community, thus enriching the data's exploitability for current research challenges.

Technically speaking, our platform is made up of two main software components: *(i)* a dataset manager and *(ii)* an interactive analytics manager. The dataset manager is responsible for handling all aspects related to datasets, spanning from their creation to their possible publication on the site. This includes various tasks such as dataset organization, maintenance, and ensuring seamless accessibility for users. The interactive analytics manager is dedicated to the computation of hypergraph statistics, dataset search, and comparison. This component provides users with robust capabilities for exploring and deriving insights from the datasets through interactive and analytical features. Together, these components synergize to offer a comprehensive and user-friendly environment for managing, analyzing, and sharing hypergraph data on our platform.

In the following, we detail our platform, its underlying hypergraph model (see Sect. 3.1), and its features, with a particular emphasis on the dataset creation and uploading pipeline (see Sect. 3.2). Additionally, we delve into the currently accessible analytics functionalities (see Sect. 3.3). More technical details about how to contribute can be found in the repositories of the platform[1], the hypergraph database[2], or on the website FAQ page[3].

3.1   Hypergraph Representations in HypergraphRepository

Under the hood, HypergraphRepository exploits the SimpleHypergraphs.jl Julia library to represent and analyze the hypergraph datasets [2,4]. This library represents a hypergraph $H = (V, E)$ as an $n \times k$ matrix, where n is the number of vertices and k is the number of hyperedges. Vertices and hyperedges are uniquely identified by progressive integer ids, corresponding to rows $(1, \ldots, n)$ and columns $(1, \ldots, k)$, respectively. Each position (i, j) of the matrix denotes the weight of the vertex i within the hyperedge j. SimpleHypergraphs.jl also provides several constructors for defining meta-information type and enables the attachment of meta-data values of arbitrary type to both vertices and hyperedges. In this manner, the library naturally models a wide range of hypergraph types, such as heterogeneous, weighted, attributed, and temporal hypergraphs.

Currently, SimpleHypergraphs.jl offers two mechanisms to load and save a hypergraph from or to a stream. In our framework, we use the plain text storage type (named HGF format). In this format, the first line consists of two integers n and k, representing the number of vertices and the number of hyperedges of H, respectively. The following k rows describe the structure of H; specifically, each line represents a hyperedge as a list of all vertex-weight pairs within that hyperedge. While this format's simplicity facilitates seamless interoperability among hypergraph software libraries, its drawback lies in the necessity to define supplementary information (such as vertex metadata or hyperedge weights) in separate files. Still, it is worth noting that the framework-agnostic nature of the HGF format means it is not bound to the development of any specific library, including SimpleHypergraphs.jl, in our case. For instance, this flexibility allows users to upload directed hypergraph datasets, even though SimpleHypergraphs.jl currently lacks support for these structures.

---

[1] github.com/HypergraphRepository/website.
[2] github.com/HypergraphRepository/datasets.
[3] hypergraphrepository.di.unisa.it/f-a-q.

3.2   A Community-Driven Hypergraph Collection

As previously discussed, our project's primary aim is to provide the research community with a comprehensive and expansive hypergraph collection that can evolve with the community's interests. This approach enables users to actively contribute by uploading their own datasets. Figure 1 summarizes the dataset management pipeline.

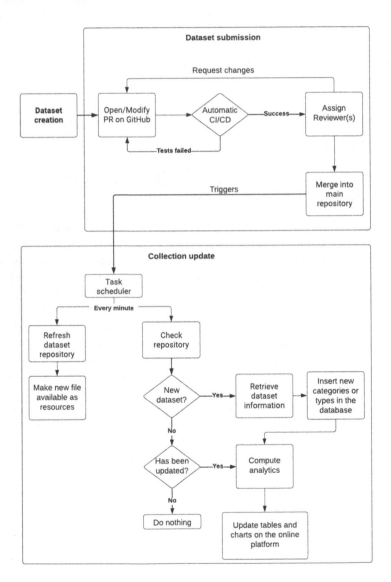

Fig. 1. Workflow of the dataset management pipeline.

Given that the dataset repository, storing all the hypergraphs accessible on the platform, is publicly accessible, any user can propose modifications. To facilitate this collaborative effort, we established a designated template to guide the process of uploading a new dataset into the repository. The overall pipeline comprises three main phases, described below.

Dataset creation phase. The first step of the pipeline is delegated to the end user who intends to upload a novel dataset. The same process applies if a user wants to modify existing hypergraph data. Upon forking the current dataset repository, the user has to create a new folder for each additional dataset they wish to contribute. This folder must include the hypergraph stored in the HGF format, a markdown-formatted file offering a concise dataset description along with any supplementary information the user deems pertinent, and a metadata file describing the hypergraph's type (e.g., homogeneous, temporal) and its application domain (e.g., social network, infrastructure network). These three files are essential for initiating a successful pull request (PR) on the main repository. Moreover, the user has the option to upload further files describing hypergraph features, such as node and vertex labels, timestamps, and hyperedge weights.

Dataset submission. Once the user is satisfied with the dataset and any supplementary data generated, they can submit all components to the central dataset repository. Specifically, the user must initiate a PR via the GitHub system, adhering to the provided template that outlines a checklist to be fulfilled before officially requesting a merge. Figure 2 illustrates the template.

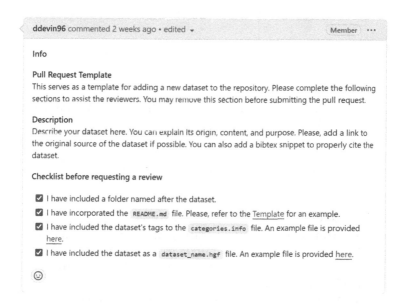

Fig. 2. PR template for adding or modifying a dataset.

Upon opening the PR, the first round of Continuous Integration/Continuous Development (CI/CD) starts. This automated process allows for validating the PR integrity before assigning one or more reviewers. This mechanism is handled through GitHub actions, which trigger specific Python scripts. The first action verifies the presence of all requested files and ensures that their format aligns with expectations (e.g., the hypergraph data adheres to the HGF format). If any checks fail, the action will halt, and the bot will promptly report the missing elements (as a comment on the PR), delineating necessary revisions. If all these conditions are satisfied, two additional actions come into play: assigning the appropriate label(s) to the PR (e.g., creation, change, documentation) and the subsequent updating of the PR after each new commit. Simultaneously, the final action, responsible for assigning a reviewer to the PR, is triggered. Each reviewer is chosen randomly from a list of volunteers, making this step a highly community-driven process open to collaboration from anyone interested. At this point, the assigned reviewer(s) can verify whether all uploaded files are compliant with the guidelines, and eventually merge the new/modified dataset into the main repository. Taking place on the GitHub platform, it is worth highlighting that the overall review process is public and unblinded. Figure 3 shows the GitHub actions pipeline.

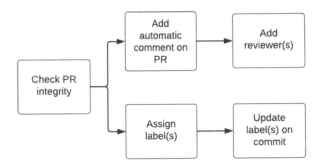

Fig. 3. CI/CD pipeline for validating PR integrity during dataset addition or modification, along with the assignment of at least one reviewer.

Collection update. The final step of the pipeline is entirely automated and is once again managed through CI/CD automation. This phase starts upon the successful merging of the PR into the main repository. Specifically, the server is configured to execute two primary scripts every minute.

- A Bash script triggers a pull operation on the dataset repository, capturing every change made. Then, it updates a folder associated with the public online collection, ensuring that all dataset files are promptly updated and made accessible for download.
- A Python script orchestrates all system calls, interacts with the GitHub APIs, and updates the database. In particular, this script makes an external call to a Julia script responsible for computing all metrics, statistics,

and plots for the newly added datasets. Leveraging the SimpleHypergraphs.jl library, the Julia script operates behind the scenes, enabling the generation, manipulation, and analysis of hypergraphs at runtime. If an existing hypergraph has been modified, the script verifies which properties have changed and subsequently updates all fields associated with those changes. To accomplish this, we utilize the GitHub API to retrieve the commit timestamps and verify which files have been updated. The coordination between these two scripts ensures the seamless update of newly computed values on the platform's website.

3.3 An Interactive Hypergraph Repository

This section overviews the features currently available on the platform. The design of these functionalities draws inspiration from established graph repositories to provide users with a user-friendly and intuitive experience as they interact with the web interface. At the time of writing, the web platform hosts two main web pages. Table 1 summarizes the statistics currently available on the website.

Table 1. Currently available statistics.

Symbols	Description
$|V|$	Number of nodes
$|E|$	Number of hyperedges
d_{max}	Maximum node degree
d_{avg}	Average node degree
d_{median}	Median node degree
e_{max}	Maximum hyperedge size
e_{avg}	Average hyperedge size
e_{median}	Median hyperedge size

– The homepage provides users with an overview of the datasets accessible on the platform, including key information such as the number of available hypergraphs, the variety of hypergraph types and categories, and a summary table detailing the notation conventions in use. Additionally, it lists the latest updated hypergraphs. As shown in Fig. 4a, users can access details about dataset names, authors, categories, types, and the number of nodes and hyperedges. The table also includes timestamps indicating when each dataset was initially uploaded and when it was last modified. Finally, for a more in-depth exploration of a specific hypergraph, users can click on the eye icon to access its specific page.

Latest updated graphs

Name	Author	Category	Type	|V| ⌄	|E| ⌄	Updated at ⌄	Created at ⌄	
algebra	ddevin96	Collaboration network	undirected	423	1,268	Jan 5, 2024 15:43:50	Dec 20, 2023 10:50:03	⊙
amazon	ddevin96	Online reviews	undirected	5,000	1,176	Jan 5, 2024 15:43:50	Dec 20, 2023 11:36:09	⊙
dblp	ddevin96	Collaboration network	undirected	71,116	25,624	Jan 5, 2024 15:57:00	Dec 20, 2023 11:36:09	⊙
email-Enron	ddevin96	Email network	undirected	2,807	5,000	Jan 5, 2024 15:57:00	Dec 20, 2023 11:36:09	⊙
email-W3C	ddevin96	Email network	undirected	5,601	6,000	Jan 5, 2024 15:43:50	Dec 20, 2023 11:36:09	⊙

Per page 5 ⌄   Next

(a) Partial view of the HypergraphRepository's homepage.

Name	Type	|V| ⌄	|E| ⌄	d_{max} ⌄	e_{max} ⌄	d_{avg} ⌄	e_{avg} ⌄	d_{med} ⌄	e_{med} ⌄	
algebra	undirected	423	1,268	375	107	19.532	6.516	10	4	⊙ View ⤓ Download (0.07 MB)
amazon	undirected	5,000	1,176	4	6	1.022	4.347	1	6	⊙ View ⤓ Download (0.05 MB)
dblp	undirected	71,116	25,624	25	69	1.244	3.452	1	3	⊙ View ⤓ Download (0.96 MB)
email-Enron	undirected	2,807	5,000	786	25	7.662	4.301	2	2	⊙ View ⤓ Download (0.2 MB)
email-W3C	undirected	5,601	6,000	282	23	2.385	2.227	1	2	⊙ View ⤓ Download (0.12 MB)

Showing 1 to 5 of 16 results   Per page 5 ⌄   1 2 3 4 ›

(b) Partial view of the HypergraphRepository's dataset page.

Fig. 4. The HypergraphRepository's website.

– The dataset page showcases all hypergraphs within the collection, as shown in Fig. 4b. Beyond the information available on the homepage, this table provides additional insights, such as the maximum, average, and median node degree, as well as the maximum, average, and median hyperedge size. Users can download a specific hypergraph of interest directly from this page, with the file size conveniently displayed for reference.

Interactive Hypergraph Search and Comparison. An essential feature for any online data repository is the provision of two core functionalities: the ability to search for specific datasets and the capability to compare them. In the following, we elaborate on the implementation of these features within our platform.

Regardless of the page users visit, they can search for a specific dataset through the tables presented in Figs. 4a and 4b. More precisely, users can search for a hypergraph based on its name, author, category, or type. Additionally, while navigating the dataset page, users can group datasets based on author, application domain (i.e., category), or type. Further, users can apply filters to hypergraph types and categories and define minimum thresholds for the number of nodes or hyperedges. As illustrated in Fig. 5, users can customize the settings of the main table to identify the most relevant datasets for their specific use cases. All numeric values are sortable in the table view, enhancing the user's ability to navigate and explore the datasets efficiently.

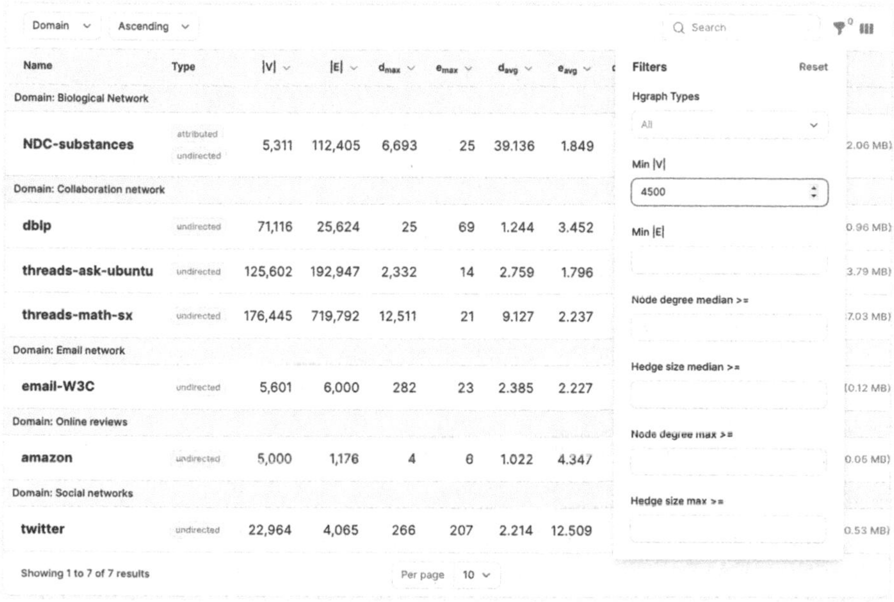

Fig. 5. Example of *filtering* and *group by* operations.

The tables accessible on both the home and dataset pages facilitate seamless dataset comparison for users. Leveraging the table headers introduced earlier (see Fig. 4), users can effortlessly compare the datasets within the collection along the dimensions defined. Although trivial, this approach offers users a bird's-eye view of the primary characteristics of the datasets. A more in-depth comparison across datasets is the object of future work.

Interactive Hypergraph Analytics. Users can access a detailed view of each dataset by clicking on the "view" button, as illustrated in Fig. 6. Currently, the user may access two different tabs reporting information about the dataset.

– The first tab reports all statistics computed for the given hypergraph, details about the dataset creator, and the dates of both the upload and the most

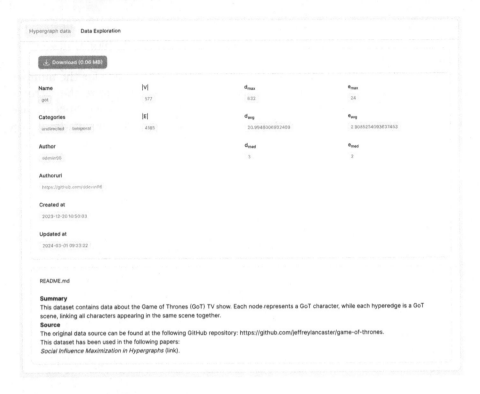

Fig. 6. A snapshot of the web page reporting the details of a hypergraph.

recent modification (retrieved through the GitHub API). This page also renders the README file with all the information provided at the time of the submission. Users are also provided with the option to download the dataset directly from this page.

– The second tab collects the interactive visualizations, providing users with the ability to visualize various hypergraph properties. This functionality is achieved by exploiting Chart.js[4], a widely acclaimed and highly customizable JavaScript library for data plotting. The incorporation of this library is designed to facilitate the development of community-driven custom plugins and the integration of features into our platform. Currently, our platform supports the visualization of node degree and hyperedge size distribution, visualized via a histogram and a scatter plot on a log-log scale, as shown in Fig. 7.

---

[4] https://www.chartjs.org/.

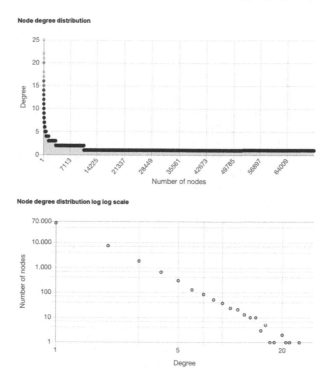

Fig. 7. Examples of node-related interactive charts.

4 Conclusion

This work presented HypergraphRepository, the first open-source, community-driven hypergraph repository. Implemented as a web-based platform, our project aims to serve as a centralized hub for hypergraph datasets, fostering collaboration, data quality assurance, and knowledge exchange within the hypergraph research community. HypergraphRepository is in its infancy, and our ongoing efforts focus on expanding the hypergraph collection in terms of both diversity in the nature and size of hypergraphs. In future updates, we also intend to:

- Enhance the set of statistics describing each hypergraph, such as adding descriptors for directed hypergraphs, information on the community composition, clustering coefficient, and high-order motifs [20];
- Offer users the option to select specific subsets of hypergraphs and directly compare them on the platform;
- Introduce interactive and scalable visualization features [14];
- Offer programmatic API access to the hypergraph database.

Acknowledgment. This work has been partially supported by the spoke "FutureHPC & BigData" of the ICSC - Centro Nazionale di Ricerca in High-Performance Computing, Big Data and Quantum Computing funded by European Union - NextGenerationEU.

References

1. Amazon, M.: GraphChallenge - Data sets (2017). https://graphchallenge.mit.edu/data-sets. Accessed 4 Jan 2024
2. Antelmi, A., et al.: Analyzing, exploring, and visualizing complex networks via hypergraphs using SimpleHypergraphs.jl. Internet Math. **1**, 1–32 (2020). https://doi.org/10.24166/im.01.2020
3. Antelmi, A., Cordasco, G., Polato, M., Scarano, V., Spagnuolo, C., Yang, D.: A survey on hypergraph representation learning. ACM Comput. Surv. **56**(1), 1–38 (2023). https://doi.org/10.1145/3605776
4. Antelmi, A., et al.: SimpleHypergraphs.jl—novel software framework for modelling and analysis of hypergraphs. In: Avrachenkov, K., Prałat, P., Ye, N. (eds.) WAW 2019. LNCS, vol. 11631, pp. 115–129. Springer, Cham (2019). https://doi.org/10.1007/978-3-030-25070-6_9
5. de Arruda, G.F., Petri, G., Moreno, Y.: Social contagion models on hypergraphs. Phys. Rev. Res. **2**, 023032 (2020). https://doi.org/10.1103/PhysRevResearch.2.023032
6. Battiston, F., et al.: Networks beyond pairwise interactions: structure and dynamics. Phys. Rep. **874**, 1–92 (2020). https://doi.org/10.1016/j.physrep.2020.05.004
7. Benson, A.R.: Data! (2021) https://www.cs.cornell.edu/~arb/data/. Accessed 30 Dec 2023
8. Bodó, Á., Katona, G.Y., Simon, P.L.: SIS epidemic propagation on hypergraphs. Bull. Math. Biol. **78**, 713–735 (2016). https://doi.org/10.1007/s11538-016-0158-0
9. Boldi, P., Rosa, M., Santini, M., Vigna, S.: Layered label propagation: a multiresolution coordinate-free ordering for compressing social networks. In: Proceedings of the 20th International Conference on World Wide Web, pp. 587–596. WWW 2011, Association for Computing Machinery, New York, NY, USA (2011). https://doi.org/10.1145/1963405.1963488
10. Boldi, P., Vigna, S.: The webgraph framework I: compression techniques. In: Proceedings of the 13th International Conference on World Wide Web, pp. 595–602. WWW 2004, Association for Computing Machinery, New York, NY, USA (2004). https://doi.org/10.1145/988672.988752
11. Bretto, A.: Hypergraph Theory: An Introduction. Springer (2013). https://doi.org/10.1007/978-3-319-00080-0
12. Commission, E.: Open science (2020). https://research-and-innovation.ec.europa.eu/strategy/strategy-2020-2024/our-digital-future/open-science_en. Accessed 29 Dec 2023
13. Fey, M., Lenssen, J.: Fast graph representation learning with PyTorch geometric. CoRR **abs/1903.02428** (2019), workshop paper at ICLR 2019
14. Fischer, M., Frings, A., Keim, D., Seebacher, D.: Towards a survey on static and dynamic hypergraph visualizations. In: 2021 IEEE Visualization Conference (VIS), pp. 81–85 (2021). https://doi.org/10.1109/VIS49827.2021.9623305
15. Hu, W., et al.: Open graph benchmark: datasets for machine learning on graphs. In: Advances in Neural Information Processing Systems, vol. 33, pp. 22118–22133. Curran Associates, Inc. (2020)

16. Khatua, A., Mailthody, V.S., Taleka, B., Ma, T., Song, X., Hwu, W.: IGB: addressing the gaps in labeling, features, heterogeneity, and size of public graph datasets for deep learning research. In: Proceedings of the 29th ACM SIGKDD Conference on Knowledge Discovery and Data Mining, pp. 4284–4295. KDD 2023, Association for Computing Machinery, New York, NY, USA (2023). 10.1145/3580305.3599843

17. Kim, S., Lee, D., Kim, Y., Park, J., Hwang, T., Shin, K.: Datasets, tasks, and training methods for large-scale hypergraph learning. Data Min. Knowl. Disc. **37**(6), 2216–2254 (2023). https://doi.org/10.1007/s10618-023-00952-6

18. Kunegis, J.: KONECT: The koblenz network collection. In: Proceedings of the 22nd International Conference on World Wide Web, pp. 1343–1350. WWW 2013 Companion, Association for Computing Machinery, New York, NY, USA (2013). https://doi.org/10.1145/2487788.2488173

19. Kunegis, J.: The KONECT Project (2017). http://konect.cc/. Accessed 30 Dec 2023

20. Lee, G., Ko, J., Shin, K.: Hypergraph motifs: concepts, algorithms, and discoveries. Proc. VLDB Endow. **13**(12), 2256–2269 (2020). https://doi.org/10.14778/3407790.3407823 https://doi.org/10.14778/3407790.3407823

21. Leskovec, J., Krevl, A.: SNAP Datasets: stanford large network dataset collection (2014). http://snap.stanford.edu/data

22. Li, P.: Datasets (2020). https://sites.google.com/view/panli-purdue/datasets. Accessed 30 Dec 2023

23. Lotito, Q.F., Musciotto, F., Montresor, A., Battiston, F.: Higher-order motif analysis in hypergraphs. Commun. Phys. **5**(1), 79 (2022). https://doi.org/10.1038/s42005-022-00858-7

24. Macgregor, P.: Datasets (2021). https://pmacg.io/datasets.html. Accessed 30 Dec 2023

25. Papachristou, M.: Datasets (2020). https://papachristoumarios.github.io/datasets/. Accessed 30 Dec 2023

26. Rossi, R., Ahmed, N.: The network data repository with interactive graph analytics and visualization. In: AAAI (2015). https://networkrepository.com

27. Sahasrabuddhe, R., Neuhauser, L., Lambiotte, R.: Modelling non-linear consensus dynamics on hypergraphs. J. Phys. Complexity **2**(2), 025006 (2021). https://doi.org/10.1088/2632-072X/abcea3

28. Wang, R., Li, Y., Lin, S., Xie, H., Xu, Y., Lui, J.C.S.: On modeling influence maximization in social activity networks under general settings. ACM Trans. Knowl. Disc. Data **15**(6), 1–28 (2021). https://doi.org/10.1145/3451218

29. Zheng, D., Wang, M., Gan, Q., Song, X., Zhang, Z., Karypis, G.: Scalable graph neural networks with deep graph library. In: Proceedings of the 14th ACM International Conference on Web Search and Data Mining, pp. 1141–1142. WSDM 2021, Association for Computing Machinery, New York, NY, USA (2021). https://doi.org/10.1145/3437963.3441663

Clique Counts for Network Similarity

Anthony Bonato$^{(\boxtimes)}$ and Zhiyuan Zhang

Toronto Metropolitan University, Toronto, Canada
abonato@torontomu.ca

Abstract. Counts of small subgraphs, or graphlet counts, are widely applicable to measure graph similarity. Computing graphlet counts can be computationally expensive and may pose obstacles in network analysis. We study the role of cliques in graphlet counts as a method for graph similarity in social networks. Higher-order clustering coefficients and the Pivoter algorithm for exact clique counts are employed.

1 Introduction

Graph similarity is a central topic in the ever-expanding, interdisciplinary field of complex networks. Quantifying similarity between networks is crucial for revealing their latent structures and in model selection. On the applied side, graph similarity is widely used in many areas, such as in recommender systems in social network analysis, accelerating drug discovery, and understanding the structural similarity of biological molecules.

One approach to graph similarity is to use counts of small subgraphs as a measure of similarity. Small subgraphs are also called *graphlets* or *motifs*. Graphlets have found wide application in many fields, such as the biological sciences, social network analysis, and character networks, often coupled with machine learning paradigms; see, for example, [1, 2, 7, 8, 11, 15, 19, 22, 23, 26, 28]. A challenge with graphlet counts is that they are often computationally expensive to compute exactly, especially for large networks.

We consider graph similarity via counts of cliques in networks. The present work also emphasizes the roles of cliques in complex networks and uses their counts as a measure of similarity. Cliques are simplified representations of highly interconnected structures in networks. For example, a clique in the Instagram social network consists of accounts linked via friendship or mutual interests. A recent model considered an evolving network model defined by growing cliques; the model simulated many properties found in social networks, such as densification power laws and the small world property; see [4]. Another recent model [9] proposed a new distribution-free model for social networks based on cliques. Cliques have also been studied from the point of view of their densification in evolving networks; see [18].

While cliques are pervasive in networks, we expect to find fewer of them in sparse networks. As such, our work is only broadly applicable to real-world

Research supported by a grant of the first author from NSERC.

networks with many cliques. In studies such as [13], social networks were found to densify and have rich community structure. One consequence is that social networks have dense subgraphs (cliques or cliques missing a small number of edges) corresponding to small communities.

Our main goal in the present paper is to show that clique counts perform as reliably as other, more sophisticated graph similarity measures in certain networks, such as social networks. Recent work by Jain and Seshadhri [10] proposed the Pivoter algorithm for exact clique count, which is well-suited for our empirical study. Studies such as [18] have continued this work. As referenced earlier, an advantage of using clique counts is that they are computationally inexpensive.

The paper is organized as follows. We define clique profiles in Sect. 2, providing definitions and relevant notation. We also consider a higher-order version of the clustering coefficient, which is another feature of our graph similarity approach. Section 3 focuses on our methods using clique profiles as measures of graph similarity. We analyze data sets from various domains and show that clique profiles perform as well as existing graph similarity methods in several cases. We conclude with a summary of the work and various open problems.

All graphs we consider are undirected. The clique (or complete graph) of order n is denoted K_n. If S is a set of nodes in a graph G, then we write $G[S]$ for the subgraph induced by S. For a node v, the *neighborhood of v* is $N(v)$ and $\deg(v) = |N(v)|$ is the *degree of v*. For further background on complex networks, see [3]; for more background on graph theory, see [25].

2   Clique Profiles

The clique profile of a graph (defined precisely below) is a normalized vector that consists of clique counts of various orders. We will use graph datasets with labels, compute their clique profiles, and examine the classification ability of clique profiles to separate labels. We also observe the change of the k-clustering coefficient, first defined in [27], on various growing networks.

We start by defining notation. Let $\mathcal{G}_k$ be the set of all non-isomorphic graphs of order k, where the graphs are arbitrarily indexed. The *count* of $G_i \in \mathcal{G}_k$ is the number of subsets $S \subseteq V$ so that $G[S]$ is isomorphic to G_i. The *graph k-profile* of a graph G is the set of the relative frequency among the counts of isomorphism-types of graphs of order k that are subgraphs of G; it may be viewed as the embedding of G into vector space, where the i-th coordinate is $g_i/(\sum_{G_j \in \mathcal{G}} g_j)$. For example, the space is 4-dimensional if $k = 3$ and 11-dimensional if $k = 4$.

Graph profiles have appeared in many works; see, for example, [11,22,26]. In [2], Bonato et al. applied the graph profiles to select which random graph model best fits character networks from novels. In [6], graph profiles are applied to determine the dimensionality of networks from certain random graph models. Computing the counts of all graphlets remains expensive, and only inexact counts of a small percentage of a graph are feasible; see [20].

For integers $3 \leq j \leq k$, let $C_j(G)$ denote the number of j-cliques in G and let $\boldsymbol{C}_k(G)$ be the corresponding $(k-2)$-dimensional vector, where the $(j-2)$-th component is $C_j(G)$. The *k-clique profile of* G, denoted $\mathcal{C}_k(G)$, is the normalized vector with each j-th component being

$$\frac{C_j(G)}{||\mathcal{C}_k(G)||},$$

where $|| \cdot ||$ denotes the 2-norm. For triangle-free graphs, we define its k-clique profile as the zero vector.

The *clustering coefficient* is a fundamental measurement in network analysis, defined in [24]; see [3] for further background. Fix $v \in V$, and let E_v be the edge set of the induced subgraph $G[N(v)]$. The *local clustering coefficient* of v is defined as

$$c(v) = c_G(v) = \frac{2|E_v|}{\deg(v)(\deg(v) - 1)} = \frac{|E_v|}{\binom{\deg(v)}{2}}.$$

The *average clustering* of a graph G is

$$\mathrm{acc}(G) = \frac{1}{n} \sum_{v \in V} c(v).$$

The *global clustering coefficient* of a graph G is

$$\mathrm{cc}(G) = \frac{\text{number of triangles in } G}{\text{number of paths of length 2 in } G} = \frac{\sum_{v \in V} |E_v|}{\sum_{v \in V} \binom{\deg(v)}{2}}.$$

The paths are not necessarily induced. The clustering coefficient of a node may be viewed as the probability of two neighbors of a node such that they are adjacent, and the clustering coefficient of a graph is the ratio between the number of triangles and the number of all triplets. An alternative way to describe $c_G(v)$ is the proportion of 2-cliques (or edges) that involve v that form a 3-clique (Fig. 1).

Yin et al. [27] generalized this idea of clustering coefficients to cliques, which we describe next; see also [12]. Let $C_k(G; v)$ denote the number of k-cliques that involve node v. Consider the case where $G = K_n$, the complete graph of order $n > k$, and fix a node $v \in V(K_n)$. Note that the subgraph $K_n[V(K) \cup \{v'\}]$ forms a k-clique, for every $(k - 1)$-cliques K of G that includes v and some $v' \in N(v) \setminus V(K)$. There are $k - 1$ ways to form K; we then have that

$$C_k(K_n; v) = C_{k-1}(K_n; v)(\deg(v) - ((k - 1) - 1))/(k - 1).$$

We also can observe that for a graph G that

$$C_k(G; v) \le \frac{C_{k-1}(G; v)(\deg(v) - k + 2)}{k - 1}.$$

Hence, we have the ratio

$$\mu_k(G; v) = \frac{C_k(G; v)(k - 1)}{C_{k-1}(G; v)(\deg(v) - k + 2)} \le 1.$$

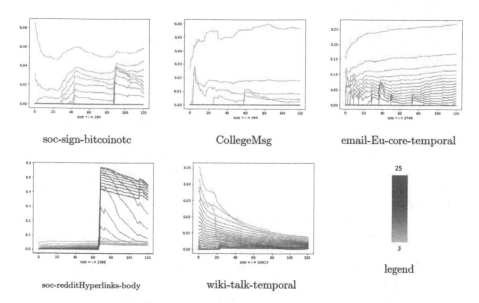

soc-sign-bitcoinotc   CollegeMsg   email-Eu-core-temporal

soc-redditHyperlinks-body   wiki-talk-temporal

legend

Fig. 1. The change of the k-clustering coefficients of five networks retrieved from SNAP [14]. For each network, every edge is associated with a time-stamp that indicates the time they arrived. We sort the edges and treat each large network as a sequence of 120 evolving networks with equal-size edge increments; every network has a fixed amount of edge growth from the previous network. We first order the edges into a list according to their time-stamps and divide them into 120 equal-sized portions. We then treat each large network as a sequence of 120 evolving networks with equal-size edge increments; that is, we use the first i parts in the list to generate the i-th network in the sequence, $1 \leq i \leq 120$. The horizontal axis indicates an index i, and the vertical axis indicates the k-clustering coefficient of the i-th network in the sequence. The label also indicates the size increment. We then record their k-clustering coefficients up to $k = 25$. The vertical axis indicates the k-clustering coefficient, the horizontal axis corresponds to the i-th network in the sequence, and the label indicates the size increment.

If $C_{k-1}(G; v) = 0$, then we simply define $\mu_k(G; v) = 0$. For convenience, we refer to $\mu_k(G; v)$ as the *k-clustering coefficient* of the node v. Analogously, we can define the *k-clustering coefficient of G* as

$$\mu_k(G) = \frac{(k-1) \sum_{v \in V} C_k(G; v)}{\sum_{v \in V} C_{k-1}(G; v)(\deg(v) - k + 2)}.$$

We observe that $\mu_3(G; v) = c_G(v)$, $\mu_3(G) = cc(G)$, and $\mu_2(G; v) = 1$.

In [27], they analyzed $\mu_k(G)$ for two network models: the Erdős-Rényi binomial random graph and the small-world graphs in [24]. They reported statistics such as the joint distributions of $(\mu_2(G), \mu_3(G))$ and the ratio of the nodes that are involved in 2-, 3-, and 4-cliques for real-world networks and graphs generated by complex network models. We consider the change of k-clustering coefficients in several large networks. See Fig. 1. There is no evidence that the k-clustering coefficients would converge to zero or any other ratio; it is also not necessarily the case that $\mu_k(G)$ is lower bounded by $\mu_{k+1}(G)$, or vice versa, such as in the soc-redditHyperlinks-body dataset.

3   Experimental Design and Methods

Our experiments aim to show that clique profiles are useful for graph similarity in certain networks. Our results will not surpass state-of-the-art results but closely match various benchmarks in the literature. Note that parameters used in clique profiles are obtainable by computing node-wise clique counts; we use Pivoter [10] for this task. We compute clique counts up to order ten and use the k-clique profile to embed each graph for $4 \leq k \leq 10$. As a separate case, we also concatenate the global clustering coefficients to these clique profiles as inputs.

We investigate the clique profiles on the following datasets, obtained from [5,16]. The networks in the dataset in [16] are known as *ego-networks*; that is, networks of reasonable order sampled from a larger network. Sampling such networks usually follows the following two steps. First, sample a node in a network with some criteria (such as a label); it is usually more meaningful to sample one with a high degree number. Second, find the neighbor set of the sampled node that meets the criteria. We then have that every network is an induced subgraph comprising the sampled node and its neighbors. We briefly summarize statistics for the datasets in Table 1.

Table 1. Statistics of the datasets.

Dataset Name	Labels	# of Labels	# of Networks (per label)
COLLAB	Subjects	3	2600/775/1625
IMDB-BINARY	Movie Genres	2	500/500
IMDB-MULTI	Movie Genres	3	500/500/500
Github Stargazers	Developer Communities	2	5917/6808
Deezer Ego Nets	User Genders	2	5470/4159
Survivor and Big Brother	TV Series	2	21/37

We next describe the datasets that we used.

1. COLLAB [16]. Networks in this dataset are extracted from scientific collaboration networks. Each node represents an author of a paper, and two authors are adjacent if they coauthored a paper. The networks are sampled from papers in *High Energy Physics*, *Condensed Matter Physics*, and *Astro Physics*, and these form the labels.
2. IMDB-BINARY and IMDB-MULTI [16]. Each node is an actor, and two nodes are joined if they appear in the same movie. The label of a network is the genre of the action movie. IMDB-BINARY dataset includes *Action* and *Romance* movies; IMDB-MULTI dataset includes *Comedy*, *Romance*, and *Science Fiction* movies.

3. Github Stargazers [21]. This is a network extracted from GitHub, where each node is a user, and two users are adjacent if they follow the same project. Networks are labeled as belonging to *web development* or *machine learning* projects.

4. Deezer Ego network [21]. The networks are extracted from European Deezer users, where two users are adjacent if they follow the same artist. Each graph is sampled from users of the same gender; *male* and *female* are selected in their sampling and form labels.

5. Survivor and Big Brother [5]. We consider datasets from two social game television shows: *Survivor* and *Big Brother*. This dataset is derived from the episodes of the shows, where the players will vote to remove each other at the end of each episode. The player with the highest number of votes is removed. The co-voting network of each season for the shows forms a directed network, with nodes representing players and directed edges corresponding to votes. We simplify these networks by taking unweighted and undirected edges between players whenever there is a voting between them. Each graph has one of the two labels that indicates the shows, either Survivor or Big Brother.

To compute the classification accuracy fairly, we use 10-fold cross-validation. For each dataset, we split into 10-folds using the *stratified shuffle split* strategy; that is, each partition preserves the percentage of samples of each class. We repeat the experiments ten times and report the mean and the standard deviation of the accuracy of the resulting 100 classifiers. We follow a similar setup of evaluation method to experiment in [16]. Primarily, we experiment using C-SVM with a linear kernel and ℓ_2-penalty, and optimize the results from $C \in \{10^{-3}, 10^{-2}, \ldots, 10^3\}$. We use the term ℓ_2-penalty to keep consistency as in [17]; some literature uses the term ℓ_i-regularization, which is the same. For the feature vectors of each graph data, we compare the classification ability between using the clique profile with or without appending the global clustering coefficient of the graphs.

We report the accuracy score using linear C-SVM with ℓ_2-penalty in Table 2. We use $\mathcal{C}_k$ and $\mathcal{D}_k$ to indicate the input feature vector using $\mathcal{C}_k(G)$ and $\mathcal{D}_k(G)$, respectively, where $\mathcal{D}_k(G)$ denotes the concatenation of the global clustering coefficient to $\mathcal{C}_k(G)$.

For additional corroboration of our results, see Table 3. We also run our experiments using linear C-SVM with ℓ_1-penalty, C-SVM with an RBF kernel, and 2-layer multi-layer perceptron neural networks with the number of neurons optimized from $\{\lfloor r \cdot k \rfloor : 4 \leq k \leq 10, r \in \{0.7, 1, 1.3\}\}$, where k corresponds to the k-clique profile and r stands for a ratio. We use Scikit-Learn 1.3.0 for all classifiers mentioned above; see [17] for details of these algorithms.

We observe better accuracy in the IMDB-BINARY and IMDB-MULTI datasets than in the benchmark. Though some results show that clique counting cannot classify the labels in the dataset, results of some datasets still verify our hypothesis that clique profiles possess a strong classification ability. For the COLLAB dataset, we did not find a graphlet-based approach to compare as a benchmark following the same experiment routine. In [22], they reported the

Table 2. The *accuracy score* of the classification using linear C-SVM with ℓ_2-penalty. Each row with $\mathcal{C}_k$ indicates the results with $\mathcal{C}_k(G)$ as an input for each network G, and each row with $\mathcal{D}_k$ indicates $\mathcal{D}_k(G)$ as the input. The last row reports the classification results using graphlet kernel from [16], where N/A stands for not available, as they are not recorded in the paper. Due to the small size of the Survivor & Big Brother dataset, the standard deviations of the results tend to be larger. The bold numbers indicate examples that make $\mathcal{C}_k$ non-unimodal.

	COLLAB	IMDB-BINARY	IMDB-MULTI	Github-Stargaers	Deezer Ego Nets	Survivor & Big Brother
acc	59.98 ± 1.57	58.03 ± 4.20	38.67 ± 2.88	51.48 ± 1.13	53.59 ± 1.48	74.29 ± 7.69
cc	63.63 ± 1.92	60.55 ± 4.91	39.09 ± 2.77	55.58 ± 1.06	50.07 ± 1.56	73.28 ± 6.48
$\mathcal{C}_4$	**69.66 ± 1.76**	70.06 ± 4.57	44.53 ± 3.01	54.37 ± 0.86	51.59 ± 1.60	**53.64 ± 11.64**
$\mathcal{C}_5$	64.90 ± 1.87	**70.37 ± 4.11**	46.97 ± 3.61	54.67 ± 0.94	50.96 ± 1.45	52.82 ± 10.52
$\mathcal{C}_6$	67.35 ± 2.11	69.91 ± 4.68	**47.44 ± 3.32**	54.66 ± 0.88	52.38 ± 1.30	52.89 ± 12.31
$\mathcal{C}_7$	**69.11 ± 1.69**	70.76 ± 4.23	46.83 ± 3.45	54.64 ± 1.19	51.49 ± 1.73	50.09 ± 11.75
$\mathcal{C}_8$	68.05 ± 1.88	70.50 ± 4.02	48.22 ± 3.50	54.67 ± 0.95	51.40 ± 1.50	**52.30 ± 12.09**
$\mathcal{C}_9$	67.74 ± 2.03	**71.10 ± 4.05**	48.83 ± 3.26	54.72 ± 1.02	51.30 ± 1.56	51.19 ± 11.92
$\mathcal{C}_{10}$	68.13 ± 1.91	70.68 ± 4.26	**49.41 ± 3.77**	54.77 ± 0.91	51.40 ± 1.43	51.11 ± 11.60
$\mathcal{D}_4$	67.23 ± 2.09	69.16 ± 4.34	45.13 ± 3.85	58.23 ± 1.25	51.16 ± 1.29	86.08 ± 6.68
$\mathcal{D}_5$	68.97 ± 1.90	70.75 ± 4.02	46.71 ± 4.31	58.44 ± 1.12	50.64 ± 1.37	83.98 ± 7.80
$\mathcal{D}_6$	68.82 ± 1.93	70.46 ± 4.17	47.39 ± 3.81	58.42 ± 1.56	51.39 ± 1.38	83.50 ± 6.76
$\mathcal{D}_7$	68.60 ± 1.77	71.53 ± 4.69	47.39 ± 3.40	58.51 ± 1.38	50.98 ± 1.75	83.41 ± 6.79
$\mathcal{D}_8$	68.08 ± 2.10	71.10 ± 4.06	48.33 ± 3.79	58.46 ± 1.32	50.94 ± 1.41	83.71 ± 7.00
$\mathcal{D}_9$	68.26 ± 1.95	71.41 ± 3.92	49.02 ± 4.06	58.44 ± 1.51	51.11 ± 1.28	84.25 ± 7.33
$\mathcal{D}_{10}$	68.91 ± 2.15	71.51 ± 4.80	49.20 ± 3.51	58.49 ± 1.01	51.05 ± 1.90	84.49 ± 6.60
[16]	N/A	59.8 ± 1.1	39.5 ± 0.7	N/A	N/A	N/A

Table 3. Additional corroboration of results in Table 2. For each dataset, the second row indicates the machine for the classification, where SVM-ℓ_1, SVM-RBF, and MLP correspond to the classification results from linear C-SVM with ℓ_1-penalty, C-SVM with an RBF kernel, and multilayer perceptron neural networks.

	COLLAB			IMDB- BINARY			IMDB- MULTI		
	SVM-ℓ_1	SVM-RBF	MLP	SVM-ℓ_1	SVM-RBF	MLP	SVM-ℓ_1	SVM-RBF	MLP
$\mathcal{C}_4$	67.64 ± 1.8	64.85 ± 2.3	70.38 ± 2.0	70.24 ± 4.9	70.47 ± 4.4	62.89 ± 7.8	44.34 ± 3.0	48.11 ± 3.2	38.95 ± 2.5
$\mathcal{C}_5$	63.13 ± 2.0	66.45 ± 2.3	70.85 ± 2.4	70.22 ± 4.6	70.40 ± 4.3	64.99 ± 5.2	45.64 ± 3.1	49.79 ± 3.3	40.44 ± 6.1
$\mathcal{C}_6$	63.83 ± 1.8	67.00 ± 2.2	70.96 ± 2.0	69.75 ± 4.2	71.45 ± 4.7	66.45 ± 6.3	46.48 ± 3.2	50.28 ± 3.7	42.90 ± 6.2
$\mathcal{C}_7$	67.20 ± 2.0	66.90 ± 2.0	70.96 ± 1.8	70.84 ± 4.0	72.47 ± 4.3	67.21 ± 4.6	46.32 ± 2.6	50.19 ± 4.0	43.67 ± 5.5
$\mathcal{C}_8$	67.75 ± 1.9	67.03 ± 1.8	70.87 ± 1.7	70.57 ± 4.0	71.76 ± 4.5	67.53 ± 4.7	46.01 ± 2.8	50.44 ± 3.4	45.22 ± 5.2
$\mathcal{C}_9$	67.52 ± 2.1	66.74 ± 1.8	70.67 ± 2.1	70.41 ± 4.4	71.60 ± 4.1	67.50 ± 4.3	47.75 ± 3.1	50.18 ± 3.5	45.45 ± 5.1
$\mathcal{C}_{10}$	67.56 ± 1.9	66.70 ± 2.0	70.71 ± 1.9	70.21 ± 4.3	71.36 ± 4.1	67.37 ± 4.7	47.77 ± 2.9	49.96 ± 3.6	47.06 ± 5.4
$\mathcal{D}_4$	65.15 ± 1.9	66.85 ± 2.0	69.51 ± 3.2	69.11 ± 4.0	70.57 ± 4.0	62.28 ± 8.3	45.16 ± 3.3	49.15 ± 3.4	38.96 ± 4.2
$\mathcal{D}_5$	65.73 ± 1.8	67.29 ± 1.9	70.87 ± 2.4	69.25 ± 5.1	70.61 ± 4.2	65.16 ± 6.3	46.21 ± 3.6	49.67 ± 3.7	41.80 ± 5.9
$\mathcal{D}_6$	67.09 ± 2.1	67.57 ± 2.0	71.25 ± 2.2	70.07 ± 4.4	72.16 ± 4.4	66.41 ± 5.2	47.61 ± 3.3	50.00 ± 3.9	42.83 ± 5.2
$\mathcal{D}_7$	68.20 ± 2.2	67.74 ± 2.0	71.69 ± 1.8	71.58 ± 4.6	71.67 ± 4.2	67.05 ± 5.4	47.19 ± 3.7	49.89 ± 3.7	44.34 ± 5.3
$\mathcal{D}_8$	67.66 ± 2.0	67.78 ± 1.8	71.61 ± 2.0	71.36 ± 4.1	71.80 ± 4.3	67.39 ± 4.2	46.19 ± 3.9	49.37 ± 4.0	45.59 ± 5.9
$\mathcal{D}_9$	67.73 ± 2.1	67.94 ± 1.7	71.74 ± 2.0	70.72 ± 4.4	71.82 ± 4.0	67.26 ± 4.7	48.69 ± 3.7	50.72 ± 3.8	47.13 ± 6.0
$\mathcal{D}_{10}$	67.89 ± 1.9	67.65 ± 2.1	71.84 ± 2.1	70.89 ± 4.6	72.10 ± 4.2	67.42 ± 4.7	48.85 ± 3.7	50.32 ± 3.7	48.13 ± 5.6

accuracy on the COLLAB dataset after randomly flipping 10/20/40% of edges as a simulation of noise with an accuracy of $72.84_{\pm0.28}\%$.

We observed that increasing k in the k-clique profile does not necessarily improve accuracy, yet the accuracy results do not appear unimodal. This situation holds for both $\mathcal{C}_k$'s and $\mathcal{D}_k$'s. We may expect that appending the global clustering coefficient necessarily increases classification accuracy; however, observe that there are a few cases where the accuracy using $\mathcal{D}_k$ is lower than $\mathcal{C}_k$. For instance, $k \in \{4,7\}$ in COLLAB dataset, $k = 6$ in Deezer Ego Nets dataset, and some other cases have only a minor difference.

4   Discussion and Future Work

We introduced clique profiles as a fast, elementary measure of graph similarity. We compared clique profiles in various social networks and found them to be accurate separators in many labeled networked datasets. The advantage of using cliques versus full graph profiles or deep learning methods is that they are computationally less expensive.

Applying our approach to more social network datasets would be interesting in future work. While our methods are less applicable to sparse networks with few cliques, one direction would be to consider profiles of sparse subgraphs such as trees to measure graph similarity. Another direction would be to extend clique profiles to hypergraphs, which are useful models for higher-order structures in networks. Implementing Pivoter makes computing the node-wise (k-)clustering coefficients more feasible. Another direction is to investigate the effects of clustering coefficients on the node classification problems.

References

1. Ashford, J.R., Turner, L.D., Whitaker, R.M., Preece, A., Felmlee, D.: Understanding the characteristics of COVID-19 misinformation communities through graphlet analysis. Online Soc. Networks Media **27**, 100178 (2022)
2. D'Angelo, D.R., Bonato, A., Elenberg, E.R., Gleich, D.F., Hou, Y.: Mining and modeling character networks. In: Proceedings of Algorithms and Models for the Web Graph (2016)
3. Bonato, A.: A Course on the Web Graph. American Mathematical Society, Providence, Rhode Island (2008)
4. Bonato, A., Cushman, R., Marbach, T., Zhang, Z.: An evolving network model from clique extension. In: Proceedings of the 28th International Computing and Combinatorics Conference (2022)
5. Bonato, A., Eikmeier, N., Gleich, D.F., Malik, R.: Centrality in dynamic competition networks. In: Proceedings of Complex Networks (2019)
6. Bonato, A., et al.: Dimensionality matching of social networks using motifs and eigenvalues. PLoS ONE **9**(9), e106052 (2014)
7. Borgwardt, K., Ghisu, E., Llinares-López, F., O'Bray, L., Rieck, B.: Graph kernels: state-of-the-art and future challenges. Found. Trends Mach. Learn. **13**, 531–712 (2020)

8. Feng, B., et al.: Motif importance measurement based on multi-attribute decision. J. Complex Networks **10**, cnac023 (2022)
9. Fox, J., Roughgarden, T., Seshadhri, C., Wei, F., Wein, N.: Finding cliques in social networks: a new distribution-free model. SIAM J. Comput. **49**, 448–464 (2020)
10. Jain, S., Seshadhri, C.: The power of pivoting for exact clique counting. In: Proceedings of the 13th International Conference on Web Search and Data Mining (2020)
11. Janssen, J., Hurshman, M., Kalyaniwalla, N.: Model selection for social networks using graphlets. Internet Math. **8**, 338–363 (2012)
12. Lawford, S., Mehmeti, Y.: Cliques and a new measure of clustering: with application to US domestic airlines. Phys. A **560**, 125158 (2020)
13. Leskovec, J., Kleinberg, J., Faloutsos, C.: Graphs over time: densification laws, shrinking diameters and possible explanations. In: Proceedings of the Eleventh ACM SIGKDD International Conference on Knowledge Discovery in Data Mining (2005)
14. Leskovec, J., Krevl, A.: SNAP datasets: stanford large network dataset collection (2014). http://snap.stanford.edu/data
15. Milo, R., Shen-Orr, R.S., Itzkovitz, S., Kashtan, N., Chklovskii, D., Alon, U.: Network motifs: simple building blocks of complex networks. Science **298**, 824–827 (2002)
16. Morris, C., Kriege, N.M., Bause, F., Kersting, K., Mutzel, P., Neumann, M.: TUDataset: a collection of benchmark datasets for learning with graphs. In: Proceedings of ICML 2020 Workshop on Graph Representation Learning and Beyond (2020)
17. Pedregosa, F., et al.: Scikit-learn: machine learning in Python. J. Mach. Learn. Res. **12**, 2825–2830 (2011)
18. Pi, H., Burghardt, K., Percus, A.G., Lerman, K.: Clique densification in networks. Phys. Rev. E **107**, L042301 (2023)
19. Pržulj, N.: Biological network comparison using graphlet degree distribution. Bioinformatics **23**, e177–e183 (2007)
20. Ribeiro, P., Paredes, P., Silva, M., Aparicio, D., Silva, F.: A survey on subgraph counting: concepts, algorithms, and applications to network motifs and graphlets. ACM Comput. Surv. **54**, 1–36 (2021)
21. Rozemberczki, B., Kiss, O., Sarkar, R.: Karate Club: an API oriented open-source Python framework for unsupervised learning on graphs. In: Proceedings of the 29th ACM International Conference on Information and Knowledge Management (2020)
22. Shervashidze, N., Vishwanathan, S.V.N., Petri, T., Mehlhorn, K., Borgwardt, K.: Efficient graphlet kernels for large graph comparison. In: Proceedings of the Twelth International Conference on Artificial Intelligence and Statistics, pp. 488–495 (2009)
23. Sinha, S., Bhattacharya, S., Roy, S.: Impact of second-order network motif on online social networks. J. Supercomput. **78**, 5450–5478 (2022)
24. Watts, D.J., Strogatz, S.H.: Collective dynamics of 'small-world' networks. Nature **393**, 440–442 (1998)
25. West, D.B.: Introduction to Graph Theory, 2nd edition. Prentice Hall (2001)
26. Yanardag, P., Vishwanathan, S.V.N.: Deep graph kernels. In: Proceedings of the 21st ACM SIGKDD International Conference on Knowledge Discovery and Data Mining (2015)

27. Yin, H., Benson, A.R., Leskovec, J.: Higher-order clustering in networks. Phys. Rev. E **97**, 052306 (2018)
28. Zhao, H., Shao, C., Shi, Z., He, S., Gong, Z.: The intrinsic similarity of topological structure in biological neural networks. IEEE/ACM Trans. Comput. Biol. Bioinf. **20**, 3292–3305 (2023)

Author Index

A
Antelmi, Alessia 159
Arnosti, Nick 47

B
Barrett, Jordan 17
Bogliolo, Alessandro 130
Bonato, Anthony 115, 174

C
Chung, Fan 1
Cooper, Colin 32

D
De Vinco, Daniele 159
Dehghan, Ashkan 65

K
Kamiński, Bogumił 17, 47
Kang, Nan 32

L
Lüchtrath, Lukas 97

M
Marbach, Trent G. 115
Milne, Holden 115

M
Mishura, Teddy 115
Mönch, Christian 97

P
Prałat, Paweł 17, 47, 65

R
Radzik, Tomasz 32

S
Sadri, Hanieh 80
Sieger, Nicholas 1
Sirocchi, Christel 130
Spagnuolo, Carmine 159
Srinivasan, Venkatesh 80
Szufel, Przemysław 146

T
Théberge, François 17, 65
Thomo, Alex 80

V
Vu, Ngoc 32

Z
Zawisza, Mateusz 47
Zhang, Zhiyuan 174

M. Dewar et al. (Eds.): WAW 2024, LNCS 14671, p. 185, 2024.
https://doi.org/10.1007/978-3-031-59205-8

Printed in the United States
by Baker & Taylor Publisher Services